Natural Selections

Natural Selections

A Year of Egypt's Wildlife

Written and Illustrated by
Richard Hoath

The American University in Cairo Press

113 Sharia Kasr el Aini
Cairo, Egypt

The author and the publisher gratefully acknowledge the permission granted by *Cairo Today* to use material by Richard Hoath originally published in the magazine between September 1990 and October 1992.

Dar el Kutub No. 2812/92
ISBN 977 424 281 5

Printed in Egypt by the American University in Cairo Press

To Marion and Jonathan

CONTENTS

ACKNOWLEDGMENTS

Natural Selections owes its genesis to a column I write for the monthly magazine *Cairo Today*. The column, simply called "Nature Notes," was the original idea of the then Executive Editor Heba Saleh, and my thanks to her, and the other editors past and present, for its success. These monthly pieces, much shuffled, clipped here and augmented there, have provided the basic framework for this book.

The root of all natural history lies not in books but outside in the field, watching and observing. Much of this book is based on my own experiences in the field and I owe a huge debt to those who have helped me familiarize myself with Egypt's wildlife. Foremost among these people have been Mindy and Sherif Baha el Din. Both have contributed enormously to the conservation of Egyptian wildlife, often without proper recognition. More personally, they have added greatly to my knowledge of the birds and animals of Egypt. Ibrahim Helmy provided me with a great deal of background information about the mammals especially, and I owe Ibrahim a special debt of gratitude for giving me a copy of *The Contemporary Land Mammals of Egypt*, co-authored by him and D. Osborn. Ken Horner opened my eyes to the birds of the Fayoum and I would also like to thank Derek Russell for his company on many field trips and for playing patient devil's advocate while I battled with identification.

Many other friends have helped, perhaps unknowingly, in the writing of the columns and the preparation of the book. My special thanks go to Elsa, Anahid, and Kate for helping me stay here when it looked as though I might have to return to London. Had I left, this book would not have been written. Thanks are also due to my family for providing unquestioning support.

A book on animals would not be complete without an acknowledgment of at least one of their number. My cat Otta provided endless uncritical company in a hot summer spent preparing the text. On more than one occasion, when I left the computer in frustration or annoyance, she strode serenely and magnificently across the keyboard, wiping off text that was clearly never meant to be printed. The ultimate in natural selection.

Introduction

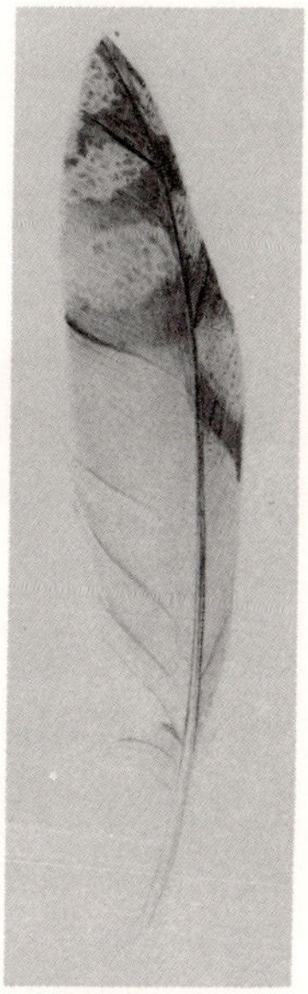

It's twilight. The air is positively chaotic with bats flitting and fluttering in a manner seemingly random yet governed by their own high-pitched sonar, sound beyond the human ear. From a branch, sparsely leafed and dusty, a pale ghost of a bird takes off on broad wings over the river. It flies so close that the white, heart-shaped facial mask and intense black eyes can be seen clearly. It circles the water and then sweeps along the banks before disappearing into the gloom. It's a Barn Owl, its potential prey one of the many rodent species now abroad but, to human eyes, invisible. In the background, above the rhythmic chirpings of a multitude of crickets, cicadas, and toads, comes a series of hysterical screamings. Their orchestrator, the Senegal Thick-knee, is out and about, but anything more than a glimpse of this nocturnal wading bird will be a bonus.

There is nothing quaint or rustic about the setting for this scene. It's the center of Cairo. To the background orchestrations one might add a cacophony of motor horns and the blaring of loudspeakers. Egypt's capital is home to some fifteen million people, buried beneath a haze of dust and pollution, and is probably one of the least prepossessing places on earth for any prospective naturalist. And yet even here natural, as well as human, dramas are played out each day. The main difference is that the natural ones for the most part go unnoticed.

Modern Egypt as a physical entity is artificial in that the lines that define it have been penned by soldiers and politicians over time rather than marked by natural geographical features. It is more or less square in shape and situated in the northeasternmost corner of Africa. To the east, and connected to it by a narrow isthmus, now breached by the Suez Canal, is an inverted triangle known as Sinai.

Egypt itself scarcely seems a promising stomping ground for the naturalist, professional or amateur. It is largely desert. The oft-quoted statistic is 96 percent desert, relieved only by the odd oasis and that green vein of life, the Nile Valley. Flowing almost due north, the Nile cleaves the desert in two, then divides north of Cairo and fingers its way into the Mediterranean through a green triangle known as the Delta. Within this 4 percent of green the majority of Egypt's population lives and most of its cities are found. It's not an overly promising geographical scenario for wildlife or for those with an interest in it.

However, there are two basic flaws in the picture as presented. Firstly, desert does not necessarily mean lack of life. While parts of the Egyptian deserts are extremely desolate and extremely arid, very few are totally devoid of life. Much of this life, particularly the animal life, may be hard to find, but it is there and is intricately adapted to living in an undeniably hostile environment. Creatures of the desert have evolved elaborate mechanisms, both physical and behavioral, to cope with these conditions, notably the lack of water and the extremes of temperature. The second flaw in the picture is the assumption that dense human presence (and population densities in the Egyptian countryside, let alone the cities, are some of the world's highest) means an absence of wildlife. Certainly, some Egyptian species have disappeared in recent years at human hands, but many remain. Some are rare and elusive. Others are common but rarely seen, and a few, like the House Sparrow and Hooded Crow, numerous and obvious. The aim of this book is to explore the variety of wildlife present in a country that may at first appear barren.

So what of this 96 percent desert? A common misconception that should be dispelled immediately is that desert is desert is desert. Visions of sand seas,

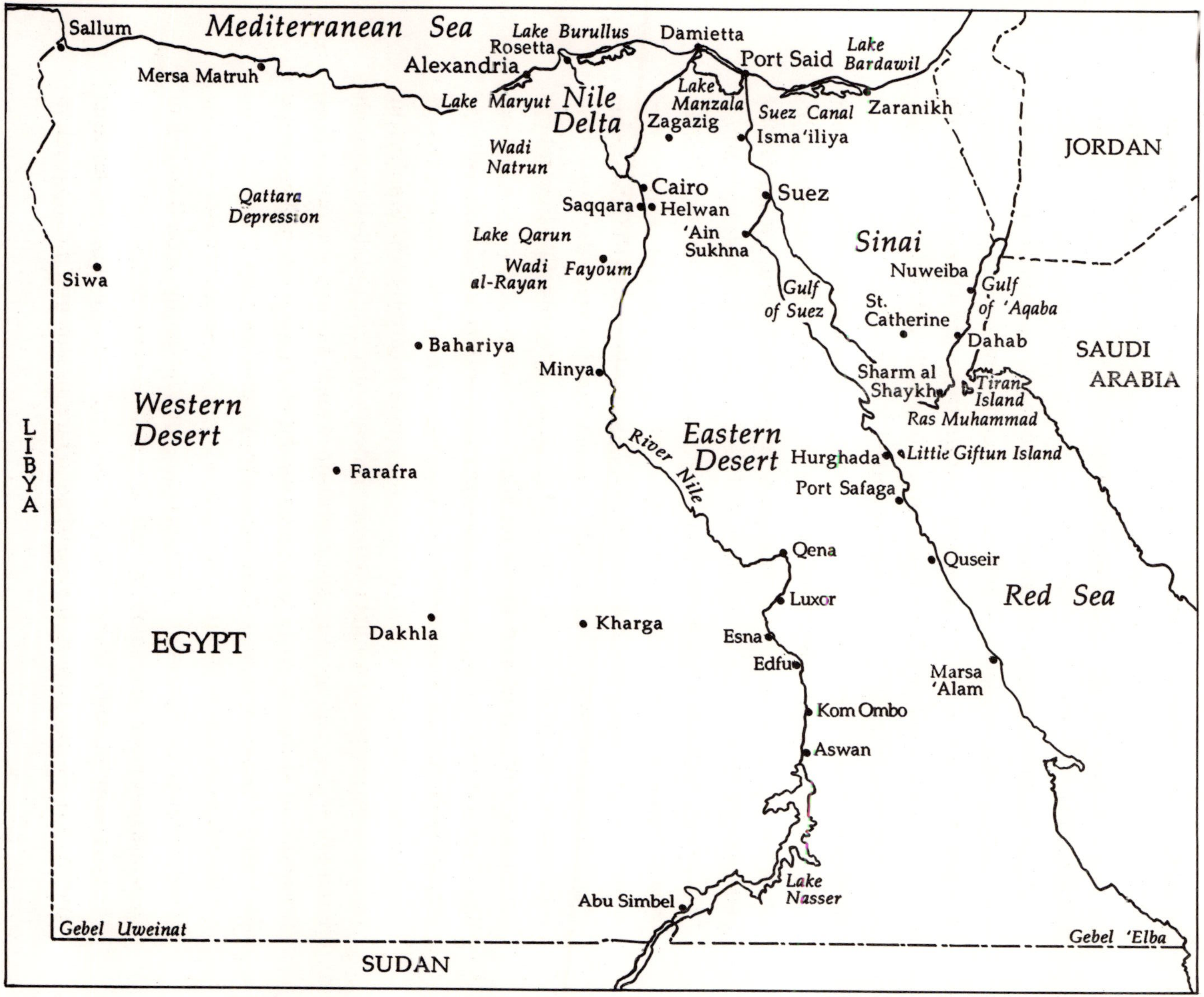
Mediterranean Sea
Sallum
Mersa Matruh
Alexandria
Lake Maryut
Rosetta
Lake Burullus
Damietta
Port Said
Lake Bardawil
Zaranikh
Lake Manzala
Nile Delta
Zagazig
Suez Canal
Isma'iliya
JORDAN
Wadi Natrun
Qattara Depression
Cairo
Saqqara
Helwan
Suez
'Ain Sukhna
Sinai
Lake Qarun
Wadi al-Rayan
Fayoum
Siwa
Nuweiba
Gulf of 'Aqaba
Gulf of Suez
St. Catherine
Dahab
SAUDI ARABIA
Bahariya
Minya
Sharm al Shaykh
Tiran Island
Ras Muhammad
Western Desert
LIBYA
Eastern Desert
River Nile
Hurghada
Little Giftun Island
Farafra
Port Safaga
Qena
Quseir
Luxor
Red Sea
Kharga
Dakhla
EGYPT
Esna
Edfu
Marsa 'Alam
Kom Ombo
Aswan
Lake Nasser
Abu Simbel
Gebel Uweinat
Gebel 'Elba
SUDAN

or ergs, of vast crescentic sand dunes are a product of romantic invention rather than topographical fact. Such areas of dunes exist, such as the Great Sand Sea west of Farafra in the Western Desert, or, far more accessible, the mini-erg just outside Cairo to the north of the Isma'iliya road, but they are far from typical. The Egyptian deserts are as varied as they are broad in expanse. Each type of desert supports a different plant and animal community uniquely adapted to its particular environment. Within this wide scope, geology and climate have conspired to create a number of broadly distinct desert zones.

To the west of the Nile lies the aptly named Western Desert. This is a vast area of flat relief. Tucked away in the southwesternmost corner of the country lies a massif known as Gebel Uweinat that just exceeds 1800 meters in height, but this is an exception. For the most part the Western Desert is relatively flat and uniform. It lacks the arid ranges and deep wadis of its eastern counterpart. Indeed, to the north lies the Qattara Depression a vast, featureless, arid expanse lying up to 44 meters below sea level.

Most people's experience of the Western Desert is limited to the new Cairo–Alexandria desert road. This speeds the motorist between the two cities, skirting the west of the Delta. The landscape is uniformly dull. Beyond the advertising billboards and ever-expanding irrigation projects lie rolling, monotonous, arid plains. To say that these roll on with equal monotony to the Libyan border would be something of an exaggeration, but by no means an extreme one. However, in many areas of the Western Desert there is sufficient vegetation to support a number of desert rodents, herds of gazelle (now much reduced), and, until recently, desert antelopes such as Oryx and Addax. The Cheetah may still exist in the remoter parts of the Western Desert. As recently as 1967 a Cheetah was shot near the Cairo–Alexandria road.

The monotony of the Western Desert is relieved by a series of oases, notably those of Bahariya, Farafra, Dakhla, Kharga, and, far to the west, Siwa. All of these are now easily accessible by road, but that does not necessarily destroy the adventure of oasis travel. Between Bahariya and Farafra lies an area known as the White Desert. It is possible to hire a guide and pickup truck in

Bahariya and head out for Farafra, spending the night in the White Desert. It is a surreal place. The desert rock, a calcareous limestone, is pure white and eroded by wind into giant, weird mushroom shapes. In the distance, to the west, rise low cliffs looking more like ice than rock in glacial white.

The trip I took was supposedly a camping trip. I had long supposed that one of the more integral pieces of equipment on a camping trip was a tent. I was proved sadly wrong. Neither tent nor sleeping bag were forthcoming, and I spent one of the coldest nights of my life amongst these wind-carved rock formations. When dawn eventually arrived my hands were so cold that I could not hold my camera still enough to take pictures. Instead I was forced to rest it on a rock, focus, and press the button. The results were, to say the least, unprofessional. In these deserts, where daytime temperatures can exceed 50°C, the nights can be close to freezing. The dramatic change in temperature from day to night can be a powerful erosional force. Rocks may expand in the heat and contract in the cold, creating and exploiting weaknesses within their structure. Furthermore, if the temperature drops to freezing at night, any water within the rock will expand, creating further stresses and contributing to the gradual attrition of the rock.

The Western Desert stretches north, but as it nears the Mediterranean coast, while it is still relatively flat and featureless, the flora and fauna change. Where vegetation was sparse it is now profuse. This is largely as a result of climatic changes. Whereas the bulk of the Western Desert is virtually rainless, the coastal strip from Sallum to Alexandria receives, courtesy of moisture-bearing winds, up to 200 millimeters of rain a year. This decreases gradually as one moves further east and dramatically as one moves south away from the coast. This narrow strip supports birds such as the Barbary Partridge and Dupont's Lark found nowhere else in Egypt, and mammals such as the Mongoose that are more usually associated with the fertile Delta.

To the east of the Nile Valley lies the equally aptly named Eastern Desert. This is very different terrain, much of it mountainous and deeply scoured by sheer-sided, flat-bottomed wadis. The hills, geologically complex and not forming one continuous range, are interrupted by plateaus of raised flat plain. Few roads traverse the Eastern Desert, but those that do give the traveler some idea of the complex topography of the region. One such road is that from Quseir on the Red Sea coast to Qus in the Nile Valley.

Quseir is a pretty fishing port still too far south to be ruined by tourist development. Its main feature is an old Arab fort with the original cannons now piled in small heaps outside the walls. From Quseir the road runs west across a narrow coastal plain into the hills. It winds its way through valleys walled by naked cliffs whose strata are rumpled and arched into huge fossil ripples hundreds of feet high. The wadi floor is very flat and evenly coated with coarse gravel. It is sparsely vegetated with tinder-dry grasses and small shrubs. On some of the peaks that rise above the road are large, regular, rectangular cairns of uncertain age or origin. Their purpose must have been navigation, for as one drives from one cairn, just before it disappears from view another becomes visible, a stone string through a desert labyrinth. The road leaves the wadi floor and passes through a region of heavily eroded, hummocky rock terrain. The rocks are flaked, peeled, and cracked and the road is forced round in a series of convolutions until the flat base of the wadi is once more rejoined. As one drives west the wadi floor widens, the great rock curtains drawing apart and the valley becoming shallower until the cliffs are distant on either side. The road now crosses a broad and near featureless stony plain until it finally descends into the contrasting fertility of the Nile Valley itself. Compared with a journey across the Western Desert one is struck by the complexity of the scenery, a product of its varied geology and erratic climate.

While wind scours the Western Desert, water is the primary agent of erosion in its eastern counterpart. Rainfall in the Eastern Desert is sporadic but, when it occurs, dramatic. No one visiting any of the eastern wadis can fail to notice its influence, the wadi floors with their sinuous meanderings and flat stony beds giving every clue to their fluvial origins. The presence of ground water along these wadis is often clear from the surprisingly rich vegetation. Gazelle haunt this desert too, but it is also the realm of the Nubian Ibex and the rare Barbary Sheep. The Muqattam Hills overlooking Cairo's Citadel represent one of the northerly extensions of the Eastern Desert ranges. Beyond, the landscape is flatter, consisting of low plateaus and rubble-strewn, rolling plains.

At the opposite end of the Eastern Desert lies Gebel 'Elba, a rock massif of great interest to zoologists and botanists alike but physically and bureaucratically inaccessible. The importance of this southeasternmost corner of the Eastern Desert is that it represents the only extension of truly African flora and fauna into Egypt. Rainfall is higher here than in the rest of the east and there is sufficient to support areas of acacia forest, many of the associated plant species being found nowhere else in Egypt. The unique flora supports a similarly unique fauna, including a number of otherwise strictly sub-Saharan butterfly species. Several birds, too, are unique to this area within Egypt, the Rosy-patched Shrike, the Shining Sunbird, the Fulvous Babbler, and the Nubian Nightjar amongst others. Sadly, however, Gebel 'Elba remains out of bounds to all but a few scientists.

The remaining area of Egyptian desert is Sinai, a peninsula that forms a land bridge between Asia and Africa. It is a bridge in the quite literal sense of the word. Animals such as the Jackal, the Striped Hyena, the peculiar Hyrax, and the flightless Ostrich spread from Africa across Sinai to Eurasia. Today this land bridge is of crucial importance in the migration of millions of birds of prey to and from their breeding grounds.

Sinai itself is topographically far from uniform. Southern Sinai is a mountainous, wadi-riddled landscape bearing much in common with the Eastern Desert. The flora and fauna, too, are similar, though the vegetation is richer, as one might expect with higher rainfall. In winter it can get extremely cold, with snow falling on the higher peaks around St. Catherine's Monastery, sometimes remaining for several months. The southern mountains of Sinai are bounded on each side by relatively barren flat plains of varying width. To the north, too, the mountains give way to sand and gravel plains interspersed with areas of dune. These stretch to the Mediterranean where they meet the coast in a series of shallow, brackish lagoons, of which Lake Bardawil is by far the largest. In winter these lakes, cut off from the sea by extensive sandbanks, are important areas for wildfowl and waders.

The Mediterranean coast is relatively uniform. From the lagoons of north Sinai one moves west to Port Said and the northern end of the Suez Canal and then on to the Delta, where the Nile, now in two main channels, meets the coast at Damietta and Rosetta. In this area, too, are important wintering areas for many species of birds—the lagoons of Lake Manzala and Lake Burullus. Beyond these is Alexandria, Egypt's second largest city and a faded shadow of its former self. Any romantic images of Alexandria are quickly dispelled on the approach from the desert road. A skyline of belching heavy industry greets the visitor, viewed across a Lake Maryut now pink with pollutants and stinking of sewage. Alexandria's back yard is well on its way to becoming an environmental disaster. Along the coast from Alexandria west to the Libyan border, vast complexes of instant holiday slums are springing up in a frenzy of tourist development. Pristine areas of Mediterranean coastline are becoming increasingly hard to find.

The Red Sea coast is very different, yet it too is threatened by development. The Red Sea is a northern extension of the huge African Rift Valley. It divides Egypt from the Arabian peninsula and, on meeting the southern point of Sinai at Ras Muhammad, is itself divided into two. The western arm forms the Gulf of Suez, culminating in the rather unattractive town of Suez and the entrance to the Suez Canal. This arm is shallow, with poorly developed coral reefs and a fauna and water quality distorted by its artificial connection with the Mediterranean via the canal. The eastern arm, the Gulf of 'Aqaba, is by contrast deep, but also warm and saline, conditions highly conducive to coral reef development. While the Gulf of Suez can be seen as a petering out of the Red Sea, the Gulf of 'Aqaba can be seen as a genuine extension of it and its faunal richness and diversity. The Red Sea itself is an extension of the Indian Ocean, and much of its animal life, vertebrate and invertebrate, bears close affinities with species of the Indian and indeed Pacific oceans. Along the Red

Sea coast and up the Gulf of 'Aqaba may be found some of the richest and most concentrated coral reefs on earth.

Corals are simple animals closely related to jellyfish. The relationship may not be immediately apparent since a lump of coral looks nothing like a jellyfish, yet they share a similar structure. A coral lump will be made up of a number of individual coral animals called polyps. Each polyp shares with the jellyfish a number of fundamental features. Both are basically circular in form, both have a single opening for absorbing food and ejecting waste, and both have a ring of tentacles around this opening armed with nematocysts, or stinging cells. There are two main types of coral, soft and hard. Soft corals lack a rigid skeleton and occur in a number of diverse forms, often with self-descriptive names such as Sea Pens, Sea Whips, and Sea Fans. A reef built of soft corals would be a pretty flimsy affair. It is the hard corals that are responsible for the coral reefs that line the Red Sea and other tropical waters. The reefs themselves represent the accumulated limy skeletons of countless millions of hard coral polyps.

Anyone who has dived or snorkeled in the Red Sea reefs, or even viewed them from the murky, scratched, and dusty hull of a glass-bottomed boat, cannot help but wonder at the dramatic contrast between the austerity of the desert and the luxurious profusion of life beneath the water surface. The key to this profusion lies in a microscopic plant or alga called zooxanthella. Each hard coral polyp plays host to many thousands of these zooxanthellae. These tiny plants, tucked away within the tissues of the polyp, are able, through photosynthesis, to provide the coral animal with oxygen for respiration and carbohydrates for food. Most importantly, and through processes not yet fully understood, they also enable the polyp to rapidly lay down a limestone skeleton. It is the accumulation of these skeletons that builds the reefs. In order for this mechanism to function properly corals require ample sunlight, and so hard corals are generally found in shallow tropical waters. No sunlight, no skeleton. No skeleton, no reef.

In their ability—courtesy of their microscopic algae—to exploit the energy of the sun, the hard corals provide the basis for the reef ecosystem. Despite their hard skeletons the polyps and associated algae provide food for a variety of fish such as butterflyfish and parrotfish, and for starfish and sea urchins. On these predators other species prey. Still more species find homes and sanctuary in the coral formations themselves. Sponges, sea squirts, and other corals use the hard coral skeletons as a base for their own colonies, and on these still other creatures prey. And so the incredibly complex mosaic of life in the coral reef is built up, its foundation an association between a tiny plant and a simple animal.

Elsewhere in Egypt the concentration of life seen in the coral reefs is only equaled along the Nile Valley, taken here to include the valley itself, the Delta, and the Fayoum. Most of Egypt's 57 million or so people live along this strip of green, either as farmers or in the cities. Too often the Egyptian countryside is described as 'timeless' or 'unchanging.' The mud-hut villages, the rural pace of life all seem from the distant past. Superficially this may be so, but even in the last decade profound and important changes have occurred along the Nile.

The most important thing about the Nile Valley as one sees it now is that it is an almost entirely human-made environment, a product of thousands of years of human manipulation, culminating in the completion of the Aswan High Dam in the early 1970s.

Pick virtually any area of the Nile north of Aswan and there will be precious little evidence of its original vegetation. The vast papyrus swamps where the pharaohs hunted hippo and crocodile have all but disappeared. At one point it was feared that papyrus was completely extinct in modern Egypt. Then, in 1968, a sad last clump of the plant was rediscovered at Wadi Natrun, an oasis on the eastern edge of the Western Desert. Today the best place to see papyrus—as plant rather than paper—is at Doctor Ragab's Pharaonic Village in Cairo. The Lotus, too, was almost lost, reduced to small colonies in the Fayoum and elsewhere.

Replacing this native flora is a human concoction based on human needs. Some of the crops grown, such as figs, dates, onions, and grapes, are native, though the varieties cultivated today differ greatly from their wild ancestors.

Many of Egypt's present crops, however, are introduced. Citrus fruits, cotton, sugarcane, bananas, tomatoes, potatoes, and Indian corn are all introduced. Even the ubiquitous fodder crop alfalfa is an introduction. With the introduced crops came associated weeds. Perhaps the most visible of these is the Water Hyacinth, which clogs up the irrigation canals and floats in great, green clumps down the river itself. Even the trees that provide shade and firewood are often introduced. Some, like the Sycamore, the Cypress, and certain willows and pines, were brought here by the ancient Egyptians. Others, such as the various eucalyptus and gum trees, are far more recent introductions.

Very little remains of the original Nilotic vegetation in Egypt. Only on a few islands immediately below the Old Dam in Aswan do remnants of the native vegetation communities exist, dominated by various species of acacia. Today even these last remnants are threatened by tourist development. As recently as 1990 the bulldozers moved onto Dun's Island south of Aswan city and bulldozed some of the few last stands of native vegetation to make way for an ill-planned hotel project. With the vegetation have gone elements of the fauna. The Sacred Ibis, the Egyptian Plover, the Darter (a sinuous, cormorant-like bird), and the Shoebill (a form of stork) have all disappeared.

Humans have attempted to master the waters of the Nile for millennia, but it was only in 1970, with the completion of the Aswan High Dam, that true mastery was achieved. Downstream, the yearly rise and fall in the level of the river has finally been controlled. Upstream, a huge artificial lake has been created known as Lake Nasser. It is beyond the scope of this introduction to discuss these changes in any detail, but certain points are interesting. With the stability of the Nile below the dam, areas that were once periodically flooded have now stabilized and been colonized by dense reedbeds. This has favored certain species of birds, such as the Clamorous Reed Warbler and the Purple Gallinule. The Pied Kingfisher, too, has expanded its range. Here, the clarity of water due to the far lower amounts of silt (now held behind the dam) is thought to be the reason. The kingfisher is finding it easier to fish.

In the Fayoum the water table has risen, resulting in the increased salinity of Lake Qarun. Freshwater fish have been replaced by saltwater species such as the Sole. Freshwater birds such as the Long-tailed Cormorant and the Marbled Teal have also vanished. However, in winter large flocks of normally coastal species of birds such as the Slender-billed Gull and the Black-necked Grebe can be found. To the south of the Fayoum a whole new wetland area has been created by excess water flowing from the Fayoum. The flooding of Wadi al-Rayan has resulted in two lakes connected by a narrow channel and Egypt's highest permanent waterfall. The reedbeds developing around these new lakes are being colonized by a number of wetland species, including the Little Bittern (a small heron) and the Little Egret.

Behind the High Dam, what was Nubia became submerged beneath the rising waters of Lake Nasser. It is one thing to subtly change the regime of a river by regulating its flow, as below the dam. It is another to simply submerge an area of several hundred square kilometers. It is doubtful whether many Nubians, much of whose history and culture disappeared with their villages beneath the lake, have much positive to say about the project. Many were displaced to the New Valley around Kharga in the Western Desert. However, Lake Nasser has benefited some animal species. The crocodile almost disappeared from Egypt but is now recolonizing areas flooded by the lake. (Tourists on cruises should note that none have been seen below the old Aswan Dam for over a hundred years.) The Egyptian Goose, something of a misnomer because it is really an oversized duck, has found the lake has created suitable breeding habitats. The new wetland areas created behind the High Dam may even lure the Sacred Ibis back to the land where it was once revered as the living image of the God of Knowledge.

If the countryside is the result of human intervention then the city is even more obviously so. Modern Cairo, the capital of Egypt, seems an extreme example of our mastery over the natural world. Anything live and green seems to be confined to balconies and traffic circles. The air smells of our effluence. Factories belch, and cars cough and splutter their way round streets either designed for a tenth their present number or laid down before the car was even invented. It seems an environment built not to merely dissuade animals and plants but to actively repel them. If that was the plan it has not worked. Even within the city center wildlife not only exists but lives, breeds, and even thrives.

Some species just cannot do without us. The Firebrat and the Silverfish are two primitive insect species that are rarely if ever found in what we might term the 'wild.' Most human city inhabitants are only too aware of the cockroaches, flies, and ants that move, unwanted, into our homes. Higher up the evolutionary ladder, Pallid Swifts have almost entirely forsaken their natural cliff-face nest sites to adopt high-rise buildings as perfectly acceptable artificial cliffs.

Many more species can do quite nicely in what we would term their 'natural' environment but are equally at home in the midst of the city. We look at a city like Cairo subjectively, applying adjectives such as ugly, pretty, messy, dirty, or picturesque to satisfy our own sense of aesthetics. To wildlife, Cairo (even though largely of human construction) is a series of habitats. A squalid heap of rubbish is generally seen by humans as unsightly, but to a mouse or rat it is a highly attractive food source with a maze of protective tunnels and plenty of suitable nest sites. Having become home to urban rodents, that same rubbish heap becomes an attractive hunting site for birds of prey, such as the Black Kite or Kestrel, or an urban carnivore such as the Weasel. An ancient tomb or modern cellar serves the same purpose to a bat as its natural cave.

Once the city is viewed as a series of habitats rather than mere urban blight then the proliferation of wildlife in and around it becomes more understandable. There are, however, degrees to which species go in adopting the city as home. Some birds such as the House Sparrow, the Hooded Crow, and the Cattle Egret make themselves obvious. Amongst the reptiles certain species of geckos, too, make themselves known. Others are more secretive and share our dwellings as largely unseen guests. The Black and Brown rats and the Spiny Mouse come into this category. It is their very elusiveness that enables them to survive. After all, who actually wants a house full of rats and mice?

As has been seen, Egypt is a far more varied land than one might expect from a country that is predominantly desert. However, it varies not just over space but over time. One need only look at the friezes in the tombs to see elaborate representations of papyrus swamps and lotus beds, habitats that have long since disappeared. It also varies in the short term. Many desert animals have had to adapt to dramatic daily fluctuations in temperature. The Fennec Fox, for instance, has developed a thick fur coat to insulate it from the cold along with complex heat loss mechanisms, foremost among which are its huge ears. Other species, particularly certain desert rodents, dig deep burrows, where the temperature is relatively constant.

Conditions also fluctuate seasonally. Resident animals change their habits and habitats in response to the changes in season, even if those changes are less noticeable than in more temperate climes. Reptiles, being cold-blooded, are generally more active in summer than in winter. Some mammals, such as certain bats, avoid even Egypt's relatively mild winters by going into a state of torpor, a kind of semi-hibernation. Others, such as the Spiny Mouse evade the cold by moving in with us. Birds, endowed with the power of flight and hence far more mobile than most creatures, reflect these changes more obviously than perhaps any other group. While many species of bird are resident in Egypt, some are summer visitors, many more winter visitors, and more still pass through as migrants on their way to or from their European breeding grounds.

It is these variations through the Egyptian year in animals' habits and in the animal populations themselves that this book follows. The species covered are merely a selection of either those most likely to be seen or those rarer species that are of special interest. There is still much to be learned about even the more common species described. It is sincerely hoped that the animals encountered in the following four seasons will act as a catalyst for the reader's own observations, opening up a whole new world.

Spring

Spring sees the natural world in a glorious state of flux. Early sunshine eases the reptiles out of their winter torpor, mammals of all kinds turn to the serious business of families, birds migrate, and insect populations explode. It is a time when even the most armchair-bound of naturalists are dusting off their binoculars, grabbing their notebooks, and heading out into the field. Non-naturalists (should such beings exist) should at least keep their eyes open.

Every spring huge numbers of birds pass through Egypt on their way from their African wintering grounds to their European and Asian breeding sites. Putting names to the immense variety of birds can be daunting, for bird identification can be a tricky business. While some species have the courtesy to announce their identity through a distinctive call or gaudy coloring, others are not so considerate. Even the experts can end up ruminating furiously over the subtle differences between a Common and a Long-legged Buzzard. They fret openly in trying to separate the sandpipers and argue at length over the sand plovers, Greater versus Lesser. But help is at hand. A friend, as yet unversed in such complexities, has come up with her own identification system. It is simplicity itself since it involves only two types of birds, differentiated by how they fly. A bird, any bird, once spotted is hailed as either a 'flapper' or a 'soarer.' The identification process thus complete, the eye is free to rove the skies for new game.

The purists may scoff, but at this time of year this simple classification is of considerable importance. There are literally hundreds of thousands of birds heading north from Africa around this time. The exact route they take through Egypt is largely dependent on whether they are 'flappers' or 'soarers.'

Many of the smaller birds, the warblers, chats, swallows, and shrikes, for instance, are active fliers. Their small size means that they need relatively low amounts of energy to sustain flight and can pretty well travel as they please. These species migrate across Egypt on a broad front, often over vast expanses of featureless desert. These are the 'flappers.'

The warblers especially are an infuriating group of birds when it comes to putting names to faces. Over thirty different warblers have been recorded from Egypt but all are generally small and dull-colored and tend to stick to thick cover. It is thus very tempting to just write them off as LBJs (Little Brown Jobs), a birdspeak term for small, dull things the observer cannot identify. However, with practice even the smallest, dullest thing can be identified, if not by sight, then by sound. While not all warblers necessarily warble, most have distinctive calls. Thus while both the Willow Warbler and the Chiffchaff are 11 centimeters long, dull olive-brown above, pale below, with a more or less distinct eye stripe, they can be readily distinguished. The former has a pleasant, melodic song described variously as 'wistful' or 'melancholy,' while the latter merely repeats its name over and over again, an incessant *chiff-chaff, chiff-chaff.* Names too can give clues. The Whitethroat is, unsurprisingly, a warbler with a white throat, a feature shared by the Lesser Whitethroat which is, once more unsurprisingly, a little smaller (though the gray rather than russet wings are a better field mark). Things can go a bit awry. A Blackcap does indeed have a black cap but only if it is male. The female is crowned with brown. The Reed Warbler is generally found in reeds but the Sedge Warbler does not need sedge, the Garden Warbler can do without gardens and the Olive-tree Warbler is quite prepared to perch in trees of various other types. Warblers are not overly worried by political geography. The Sardinian Warbler is still a Sardinian Warbler as it migrates across Egypt's deserts.

A far more striking group of birds, and an easier one for beginners to cut their ornithological teeth on, are the shrikes. Also called 'butcher birds' for their habit of impaling insect and small vertebrate prey on thorns, the shrikes are a distinctive bunch. Apart from the Rosy-patched Shrike, tucked away in the inaccessible mountains of the southeast, the only breeder is the Great Gray or Northern Shrike. In early spring the Great Gray is joined by other

Great Gray Shrike. *This juvenile is distinguished by its pale bill and rather short tail. While I was watching it skulking in the midst of a citrus tree, the parents were creating a terrific fuss trying to distract my attention. I left quickly.*

shrikes passing through on migration, the Lesser Gray, the Red-backed, the Masked, and the Woodchat. While not flamboyantly colorful, a more boldly patterned group of birds could not be wished for. Though all have a black *Lone Ranger* mask, each has its own characteristic plumage in black, gray, and white, often with bold splashes of rufous brown or chestnut. Added to this is their spotter-friendly behavior. Skulking round in dense thickets is not the shrike's style. All perch boldly and openly on such bold and open things as telegraph wires or electricity pylons.

A springtime visit to the Japanese Gardens in Helwan produced an interesting migrant one year. The gardens themselves are worth a visit. With a late rainstorm came the revelation that trees bore green rather than beige leaves and the gardens seemed to be a good place to head for. Previous visits had been disappointing. The Helwan Japanese Water Gardens are indeed in Helwan. Rows of chipped and cracked plaster Buddhas and a vaguely oriental-style tea shop grant some concession to Japan. But water? This seemingly indispensable ingredient of any water garden was conspicuous in its absence. There were empty concrete pools, and bridges over empty concrete pools, but nothing that was actually wet. Rain, I thought, might have resolved this.

In this respect I was disappointed (the concrete evidently leaks), and the day ended on something of a low note with me being chased out of the park by an irate gardener. In one of the trees that overhang the chipped Buddhas, I had spotted a Wryneck. The Wryneck is an aberrant woodpecker some 17 centimeters long and cryptically colored in beiges and vermiculate browns. It looks nothing like a woodpecker, more like a heavily camouflaged warbler, but the experts are adamant. It is rather scarce and so I attempted to get nearer. In trying to get a closer look I wandered onto a patch of dry mud sparsely prickled with yellow spikes of lifeless grass stalks. This 'lawn' turned out to be the park-keeper's pride and joy. Migrating Wrynecks are uncommon and I tried to explain what I was doing. Not knowing the Arabic for Wryneck, and he not knowing either the English or the Latin, I resorted to mimicking the neck-twisting behavior that gives the bird its name. This hardly struck a chord and

Wryneck. The delicately marked plumage of this weird woodpecker renders it almost invisible in the trees. Luckily for birdwatchers it often feeds, as here, on the ground.

after a brief spate of mad pointing I found that the bird had resumed its migration and beat a retreat.

Enough of 'flappers.' The 'soarers' include the seriously large birds, the bigger birds of prey, the pelicans, storks, and cranes. For these birds sustained active flight would require far too much energy. Instead they make use of upward spiraling currents of warm air. Armed, literally, with broad wings, often of huge span, they use these to ride the rising thermals, ascending almost effortlessly to then glide on their way. The drawback is that these thermals do not rise over open water. They must be channeled up by cliffs, crags or mountains. Thus the larger migrants are funneled down migration routes at either end of the Mediterranean, avoiding open water crossings to Europe. Many of the larger, more impressive birds are forced by aeronautic design along the Red Sea coast and on up the Sinai to the Levant. For the very largest even the Red Sea marks a formidable barrier and they cross at Suez, the first landbound crossing, where Africa meets Sinai at the head of the Gulf of Suez.

Steppe Eagle. The piercing eyes and sharply hooked bill of the Steppe Eagle are the hallmarks of a powerful predator.

Suez is not an obvious weekend destination. With its seemingly endless industrial zones, queues of smoke-belching container ships, and oil-coated shorelines it falls somewhat short of most people's idea of a dream town. However, it is one of the best places in Egypt for watching birds of prey. A spring visit may be rewarded by the sight of these raptors in their hundreds over the city. Imperial, Lesser-spotted and Steppe eagles may all be seen, but it is the Griffon Vulture that is perhaps the most impressive. Now rare on migration, the Griffon is known as the 'tea-tray in the sky,' an apt description, with its near oblong silhouette and a wingspan of some three meters.

Watching birds at Suez is unlikely to prove a quiet affair, a situation that has nothing to do with noise from the ubiquitous industrial plants. Suez is one of the few places in Egypt where it is possible to see, and virtually impossible not to hear, the Indian House Crow. If ever a bird can be said to nag then this has to be it. They seem to do little else but perch and squawk—not a mild avuncular *caw*, but an earsplitting *CAAARRRR*. Egypt is not the Indian House Crow's natural home. They are thought to have arrived from their native subcontinent on ships earlier this century. Unlike the Hooded Crow, familiar to Cairo residents, the House Crow is black with the gray restricted to the nape. And then there's that noise.

An unexpected result of the 1990–91 Gulf Crisis was an unusual encounter with one of the migrating giants. As tourism dried up, the resorts emptied and I headed off to the Sinai to find I had the peninsula almost to myself. Indeed about the only other creature staying in my hotel was a bird. Curiously, there are people who feel somewhat disconcerted in finding a meter-long vulture perched on top of their hotel rooms. I am not of that club. Indeed the more vultures, or any other creatures for that matter, the better. I was therefore delighted to be sharing hotel space with an immature Griffon Vulture. 'Immature' should not conjure images of smallness and fluffiness. Vultures mature slowly, and this was by any standards a class-one vulture, the only difference between it and an adult being the beige rather than white ruff and its black bill.

This particular bird had been found starving in the desert and reared back to health by the hotel staff. Free to go at any time, it had elected to stay where the food was regular. Later, as spring progressed, more natural migratory urges took over and the Griffon followed its peers to its Mediterranean breeding grounds, but in the few weeks this individual spent at the hotel, it managed to convert the rooftop bar into something akin to Bombay's Towers of Silence. I for one was delighted. Vultures get a pretty poor press, fueled by pictures in glorious Technicolor of ravenous, bloodied hordes ripping away at a carcass on some African plain. However, as Nature's trash men they fulfill

Indian House Crow. *Most field guides show pictures of the Indian House Crow with its bill closed. It is an incredibly raucous bird, and this is a far more typical pose.*

a valuable role, though admittedly hardly an aesthetic one. Their appearance is a result of adaptation to this role. The huge bill is designed for tearing flesh and sinew and ripping away at toughened hide. (However, it can also be used with amazing dexterity, as the hotel vulture demonstrated in systematically unweaving the straw roof of the rooftop bar.) The neck is serpentine, to reach inside a carcass, and sparsely feathered, a prudent evolutionary move since it would be hard to keep clean. The huge, broad wings enable the bird to fly vast distances in near-effortless soaring flight.

While the Griffon is now only a migrant in Egypt, three other species of vulture breed. The Egyptian Vulture, or Pharaoh's Chicken, is a scruffy white bird with a slender bill and looks nothing like a vulture. It is a rare breeder and more common migrant. The Lappet-faced and Bearded vultures are very rare residents in the Eastern Desert and, in the case of the latter, the Sinai Mountains. Both have wing spans of around three meters.

As fate would have it my trip to the Sinai coincided with the release by Saddam Hussein's troops into the Arabian Gulf of what was possibly the largest oil slick ever. As millions of gallons of black crude spewed into the ocean the inevitable pictures of oil-soaked sea birds appeared on the television screens. Footage of dying cormorants flopping helplessly in a suffocating black sludge had an added poignancy in the Sinai. The coral ecosystem is similar to that of the Gulf, the bird, reptile, and mammal species familiar. The gulls and terns we had seen that morning from the dive boat were the same species as the ones we now watched floundering in thickened crude on the TV. Birds provided the most 'newsworthy' images, but also affected were whales, dolphins, porpoises, and sea turtles. And Dugongs.

The Dugong is similar in form to a large, rotund seal. However, it has no hind limbs but a tail fin like a dolphin's. The forelimbs have become flippers. Apart from some sturdy bristles round the snout it is hairless, with leathery, gray skin, paler beneath. The eyes are tiny. This curious animal spends its time grazing beds of sea-grass and, along with the Manatee, is the only vegetarian marine mammal. Seal? Dolphin? Whale? In actual fact the Dugong's closest relatives are the elephants and the curious rabbit-like Hyrax. Fascinating though the Dugong is, not even its most ardent admirer could call it beautiful. It therefore somewhat stretches the credulity to find that the Dugong is the creature on which the mermaid legend is supposed to be based. Having teats between the front flippers, the female Dugong nurses her young much as a human mother does, holding the baby Dugong to her leathery breast. Sailors seeing suckling Dugongs and using copious amounts of imagination (or rum) considered these creatures to be mermaids or sea nymphs. Fond as I am of the Dugong, I think this must be a case of sailors being at sea far too long. Dugongs are still reported from the Saudi side of Tiran Island in the Gulf of 'Aqaba but are now at best extremely rare in Egyptian waters. One reason can clearly be seen at the Oceanographic Institute in Hurghada where there are numerous very badly stuffed specimens on exhibit as well as various black-and-white photographs of hunted Dugongs in various stages of dismemberment. With the eco-carnage in the Arabian Gulf, the Red Sea population of the vanishing mermaid takes on a new importance.

I was lucky enough to spend one early spring in the Sudan with some of the migrant birds before they decided that it was really getting a little hot down there and Europe, via Egypt, beckoned. Khartoum is a city blessed with a large area of acacia woodland known as the Sunt Forest. An amble round the Sunt Forest can be thoroughly recommended to anyone heading in that direction, though care should be taken: the equipment of the natural historian, the notebooks, binoculars, and long-lens cameras, is just that used by others in more covert and militarily sensitive forms of employment. That apart, the forest is a wonderful place. The woods are alive with such exotics as Gray and Red-billed hornbills as well as bickering gangs of White-headed Babblers. However, it was most memorable for the huge numbers of Masked Shrikes. I have seen Masked Shrikes passing through Egypt in spring but never had I seen so many in one place. It seemed that the Sunt Forest was a pre-migratory gathering point, a not unreasonable theory given its situation at the confluence of the White and Blue Nile valleys. The Masked Shrike is a striking, long-tailed

Griffon Vulture*. This is the young bird I shared a hotel with in Sinai. Vultures rely on their powerful bills to feed. Their feet are relatively weak and the talons blunter than an eagle's.*

bird, predominantly black and white with the underparts pale cream suffused with buff. The name comes from the black band through the eyes typical of the group, giving it the appearance of an avian raccoon. The wood was alive with its harsh *kreer kreer*, and a shrike seemed to be dropping from every other tree onto some insect prey.

Lying at the confluence of the two Niles, Khartoum proved excellent for observing water birds, including three species of tern that would later be appearing along the Egyptian Nile Valley. The White-winged Black Tern had yet to turn black, and neither were its wings white, but it could be identified with some care. The Whiskered Tern in early spring plumage was fairly similar but with the makings of its summer black cap. The Gull-billed Tern, a not infrequent visitor to Aswan in spring, looked significantly bigger than either. By the time they reach Aswan they have adopted a coal-black cap and this, together with the heavy black and unternlike bill makes them easy to identify.

Migration is not purely the realm of birds. Birds do it, and while bees don't, certain butterflies do. A spring excursion into the desert can find one surrounded by Painted Ladies. The Painted Lady is one of Egypt's commoner butterflies, a large but fragile-looking insect, checkered in faded browns and reds and marked with white. Its delicate appearance is deceptive, for the Painted Lady is the long-distance migrant of the insect world. Many of the individuals seen fluttering along the desert margin will be heading for Europe, finding their way as far north as the Arctic Circle. Others will remain in Egypt to breed. The caterpillars are dark and spiky and have an extraordinarily catholic taste in food plants.

The other butterfly worthy of mention is the Plain Tiger. There seems to be an unwritten rule in the natural world that the commoner an animal is the duller it is. Thus in Egypt the muted House Sparrow and the black and gray Hooded Crow are the most ubiquitous birds, and the somber House Mouse and Black and Brown rats are the most numerous mammals. But every rule has its exception and this particular rule is broken by the Plain Tiger. Not only is it the country's most common butterfly, it is happily also its most striking.

It is a robust creature as butterflies go, with a wingspan of some six centimeters, colored largely rusty orange. Dark tips marked with white decorate the forewings, and the hindwings have dark rims dotted with small white spots. It can be seen virtually anywhere, from city-center roundabout to Delta palm groves. But beware of imitations. With the Plain Tiger you must make sure you have the real thing. The Plain Tiger may look striking, but it tastes vile, or so I am told. Therefore, most insect-eating predators leave it well alone. A naive bee-eater may, for instance, catch one and discard it in disgust. The now rather less naive bee-eater will henceforth avoid butterflies of this particular color and pattern. Enter *Hypolimnas misippus* (there is no common name). This butterfly tastes delicious, though again I have taken others' words for it. The female of the species has therefore evolved to look just like the Plain Tiger, and can only be distinguished from it by the slightly larger size and the broader border to the hindwings. Predators, perhaps our slightly wiser bee-eater, seeing the distinctive pattern assume the potential meal will taste horrendous and steer well clear. The male, presumably more expendable, is of similar proportions but is uniform black with an oval white patch on each wing.

Hypolimnas is found throughout Africa, and the male seems to be a particularly belligerent butterfly. I have seen him on several occasions defending a small territory not just against others of his own kind, but against other species, even other insects. While I have not yet witnessed this behavior in Egypt, anyone savaged by a butterfly should check for dark wings and white spots. A riled *Hypolimnas* may make an unusually irksome adversary.

The Cuckoo, traditional harbinger of spring in northern climes, is but a fleeting visitor to Egypt, passing through in small numbers as it heads for Europe to become the parasitic scourge of many a small bird. It is only when the males arrive at their breeding grounds that they can sit back, draw breath, and utter a timely and onomatopoeic *cuk–oo*. In Britain the first one to do so is rewarded with a mention in the letters column of *The Times*. Another sign that warmer weather is on its way is the swallow, but in Egypt the swallow is also a common resident quite able to find enough insect food to survive the winter here. As a seasonal herald the swallow is therefore pretty second-rate, with one proviso. It is easy to tell the birds just passing through from those that are going to stick out Egypt's sticky summer heat. While both races share the deeply forked tail and steely blue upperparts, the resident swallows have deep reddish underparts. In the migrating swallows the red is restricted to the throat, the underparts being otherwise pale cream. That aside, for me spring has truly sprung when the evenings begin to warm and the air, round virtually any patch of watered green, becomes soothingly athrob with the crepuscular orchestrations of the Egyptian Square-marked Toad.

Egypt, being predominantly desert, is not graced with a wide selection of amphibians, a group for whom water is a fairly crucial requirement. Two species of frog and four species of toad is the generally accepted complement. Of the latter only two are at all widespread. The Green Toad is, by toad standards, an attractive creature in pale gray, densely marbled with bright green. The other is the much commoner and rather clumsily named Egyptian Square-marked Toad. It is a duller, browner animal than the Green Toad, though a similar marbled pattern is still discernible, especially in younger animals. The underparts are pale and there may be white spots or a white stripe down the back. I have seen many of these toads and am not entirely sure where the 'square' bit comes in, but the name remains. This is the common toad of Egypt and can be found almost anywhere that suitably damp conditions exist. The warty, slimy skin that does not endear the toad to many carries a mild irritant and is hence partly protective. It also helps prevent the toad from drying out. Toads can easily be told from frogs by this warty skin and by the fact that they walk rather than hop. Only when threatened will a toad hop. The rhythmic churring calls that fill the spring air are the nuptial serenades of courting males, vocally vying for the much larger females. If successful, the male will clamp the female in a vice-like grip, known as amplexus, during which fertilization occurs. Though toads are less tied to ponds and ditches than frogs, the eggs must still be laid in water.

Plain Tiger *and* **Hypolimnas misippus**. *The striking similarity between the Plain Tiger (above, male) and the female* Hypolimnas *(below) is no accident but a defense mechanism evolved by the latter.*

Hypolimnas misippus. *The male* Hypolimnas *bears no resemblance to the female (see previous page). This one is seen settled on its larval foodplant, the Dentated Dock.*

Toad-watching is a more fascinating and relaxing way to spend an evening than it might otherwise seem. I can remember one evening spent seated on a lawn beneath the rustling green umbrellas of Royal Palms watching a little piece of theater. The toads, full of that confidence that comes with spring, were striding across the grass with no attempt at concealment. A weasel, that wily, wiry Cairene predator, was eyeing them from the cover of a hollowed palm log. A cat was eyeing the weasel. Cat watching weasel watching toad. Then, bumbling onto the scene comes a young hominoid, a bit actor with no sense of the unfolding drama. Nothing came of nothing as the three potential antagonists scattered, the toad rather more slowly than the others, in the face of this disturbance. Had the sketch run its course, the weasel would have made short shrift of the cat but probably met his match with the toad. The toad would have inflated itself to something approaching twice its normal size to confront its attacker. If this had not put the weasel off then the foul-smelling secretions from that warty skin surely would have. Facing a toad in such a mood would be like confronting a particularly oily sausage covered in boils. But it does have a beautiful, hypnotic voice.

Other creatures, unable to serenade potential mates, resort to different strategies. Even amongst the reptiles, a group not renowned for elaborate courting, veritable dandies can be found. Take the agamas. For most of the year the agamas are robust lizards, some 30 centimeters long, rather flattened and noticeably spiny, but colored somberly in pale beiges, grays, and browns. However, come the breeding season and the males, particularly those that are dominant (for the agamas are hierarchical beasts), become quite gaudy. The throat turns bright blue, sometimes spotted or striped with white, the underside sometimes assumes a similar shade, and the sides turn violet. Looking not unlike a reptilian peacock or a psychedelic dinosaur, the male agama goes about the serious business of wooing. None of Egypt's five agama species can be regarded as a true city dweller, but the species most likely to be encountered around Cairo is the Changeable Agama. The name stems from the less than predictable color scheme of the reptile. In Sinai, the closely-

Changeable Agama*. A male agama surveys his territory. At any sign of a threat he will embark on a bout of energetic head-bobbing.*

related Sinai Agama and the Starred Agama occur. Both may be seen around the Monastery at St. Catherine. As the weather warms they can be spotted basking in the open, head raised in the air as if surveying their domains.

In the animal kingdom, the agama may be considered fairly normal. It is the male who is bigger, more brightly colored, and more belligerent. It is the male who holds and defends a territory. He may have one mate or he may have a whole harem. Having mated he may help rear the young but often he avoids such domesticity. It's an oft-repeated system. However, skulking deep in the dank vegetation of the Delta is one of the big exceptions. The Painted Snipe is something of an enigma. For a start it is not a snipe but is placed in a discrete family of its own. However, it does bear a passing physical resemblance to a snipe with its long bill, large eyes, and wader-like appearance. In habits it is more rail-like, as for instance in flight, when it rises from the marsh with its legs dangling. The Painted Snipe is fairly common but rarely seen due to its nocturnal habits and its tendency to stick to thick cover. Most remarkable, though, is the fact that the Painted Snipe reverses virtually every sexual norm. The female is slightly larger and more colorful than the male. In spring she establishes a territory, calling to attract males and pairing with up to four different mates. It is she who initiates courtship, displaying her intricately patterned wings and tail and circling the male, calling at the same time, a persistent *kok kok kok*, spurring him on to greater things. Nature still demands that the female Painted Snipe lays the eggs, and this she does in a nest built by the male. Duty done, she then goes off to find a new husband. The abandoned male is then entirely responsible for incubating the eggs and rearing the young. The female starts calling in late February, firstly at night but then, as the season progresses, also by day.

The best chance of seeing Painted Snipe is at twilight when they move out from their daytime hideouts to more open fields to feed. Gebel Asfar, just outside Cairo, seems to be a prime place for Painted Snipe. I have only once failed to find them there. Gebel Asfar is the site of a large sewage farm. Birdwatching may seem to appeal to those with a love of the great outdoors. It might be thought splendid wholesome stuff, what with plenty of fresh air and exercise. However, for the real diehards fresh air is pretty second-rate. While it has its uses in clearing the lungs of city smog and eliminating that feeling of breathing through a car exhaust pipe, for the enthusiast there is nothing like a whiff of sewage to get the adrenaline going. Sewage means water and water means birds. During migration the Painted Snipes of Gebel Asfar are joined by a multitude of other waders, while the fertile farmland around attracts large numbers of chats and warblers and multitudes of Black Kites. Constant companionship is provided by the Spur-winged Plover. The Spur-winged Plover is unusual in that it is not only common but also striking to look at, a rare combination. The upperparts are gray-brown, the head and underparts black, while it sports a white collar extending to the cheeks but not quite meeting across the chest. The wing spurs that give it its name are difficult to see in the field. Spur-winged Plovers are by no means confined to Gebel Asfar but are highly visible residents of irrigated areas throughout the Nile Delta and Valley. They are among the first birds to colonize newly irrigated areas, and in such places can often be found in small flocks gathered at the

***Painted Snipe**. A great deal of patience is usually required to see Painted Snipes. Only at dusk do they emerge, like this pair, from dense cover to feed. The slightly larger, more colorful female is seen here in front of the male.*

side of irrigation canals as if waiting for the desert to turn green. More likely, in these early stages the canal is the best source of food.

The Spur-winged Plover is highly territorial, and any intruders, human or otherwise, are informed loudly and clearly if they enter the territory of a pair of these birds. They get very agitated, circling on broad, rounded, black and white wings and calling incessantly with a loud *tick, tick*. Against birds of their own species they have a special threat display. The defender rears up facing the opponent and exposing the strikingly patterned underside of the wings, all the while ticking away. It all adds up to a high-profile bird, and Gebel Asfar has more than its share. In parts one can scarcely move without being ticked by one pair or another.

Gebel Asfar's real claim to fame, though, is the White-breasted or Smyrna Kingfisher. Nearly twice the size of the European Kingfisher, the Smyrna Kingfisher is unmistakable. It has a white breast, chocolate-brown head and belly, and cobalt-blue back and tail. The bill is vast and coral red. The Smyrna Kingfisher is a fairly recent colonizer, and Gebel Asfar is one of the very few places in Egypt where it might be seen, indeed one of the only places in Africa where this impressive bird can be found.

The Smyrna Kingfisher is not alone, for it is amongst the birds that the truly kaleidoscopic color schemes occur. As a group, the bee-eaters are amongst the most striking, and Egypt is happily graced with three species. The bee-eaters are all slim birds with long, sharp, slightly downcurved bills and with the two central tail feathers elongated. Though their diet is somewhat suggested by their name, they also feed on a variety of other insects, including wasps, butterflies, and dragonflies. Come April the first European Bee-eaters arrive in the Delta in loose flocks, circling on pointed wings and calling with a lyrical *croop croop*. Their brilliant plumage is best described as harlequin, a gaudy mixture of chestnut and iridescent blues and greens, with a brilliant yellow throat and narrow black eyestripe. Most will be passing through Egypt on their way to provide a splash of tropical color to their southern-European breeding sites, but a few remain to nest in northwest Sinai. The Blue-cheeked Bee-eater

European Bee-eaters. *A delight for birdwatchers, bee-eaters are not only brightly colored but often, as this pair, perch right out in the open.*

is of similar size and form to its European counterpart but very different in color. It is largely bright green with a similar black mask, but with pale blue cheeks and a chestnut throat. Its habits too are similar, and a flock of bee-eaters uttering incessant *croops* while silhouetted against the sun is difficult to identify. However, Blue-cheeked Bee-eaters are summer visitors to Egypt

European Bee-eater

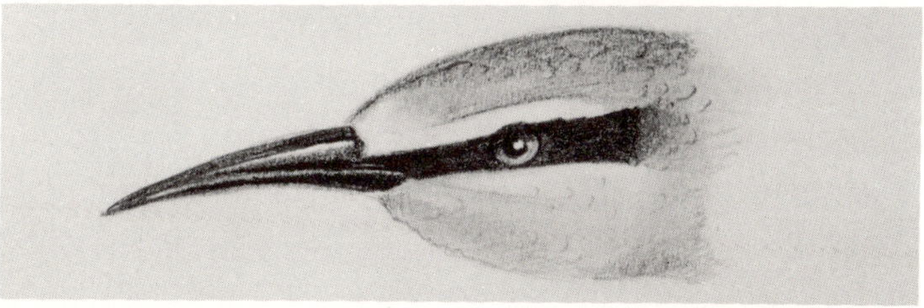

Blue-cheeked Bee-eater

and remain here to breed. The nest is a tunnel hewn by the birds in a sandbank, generally in a loose colony. It appears amongst the Blue-cheeked Bee-eater fraternity that a bank is a bank, be its origin natural or human. They are quite able to take advantage of the steep, earthen sides of the larger irrigation ditches, and in the Saqqara area, around the Sun Temple, have adopted the archaeological excavations, seemingly undisturbed by the comings and goings of archaeological excavators. The Blue-cheeked Bee-eater is most likely to be confused with Egypt's smallest and only resident member of the group, the Little Green Bee-eater. The name is a sufficiently accurate description of the bird, the green being relieved only by a narrow black mask, a black gorget, and the rufous underwings. The length of this attractive little bird is exaggerated by the two central tail feathers being rather more elongated than in the migratory species. They can be seen throughout the year along the

***Little Green Bee-eater**. Although brightly colored, this species can be almost invisible when waiting for its insect prey in thick foliage, in this case a Murray Red Gum tree.*

length of the Nile Valley, but for their bright coloration are surprisingly unobtrusive. The best way to spot them is to look for a flash of bright chestnut as they fly out from a tree or bush after some insect. The challenge is to find them once they return to their perch. Though they do not head for deep cover, locating a pair of motionless little green birds in a tree-load of motionless little green leaves is somewhat exasperating. The easiest place to find them is hawking for insects over Giza Zoo.

Another colorful springtime visitor, though once again only passing through, is the gaudy and curiously named Roller, a relative of the bee-eaters. The Roller gets its name from its spectacular, tumbling display flight. Sadly, these aerobatics are reserved for higher latitudes, but the Roller is nevertheless an eye-catching visitor. It is the size and build of a small crow but with a disproportionately large head. However, while crows are garbed in dowdy grays and blacks, Rollers play the dandy in brilliant cobalt, turquoise, and chestnut. The combination of two shades of blue led the Indians to christen them 'Oxford and Cambridge Birds'. Unfortunately rather scarce on migration, they command attention when seen, most often in early May.

However, as any naturalist who has tried to locate the brilliantly citron male Golden Oriole in thick shrubbery knows, bright colors do not necessarily mean a bird is easy to see. Both bee-eaters (with the sometime exception of the Little Green) and the Roller are model subjects though. They have the helpful habit of perching in the open on those most wonderful of inventions, telegraph wires. While fulfilling a secondary function for telephone users, the prime importance of these wires is as bird perches, for which their inventor should be saluted. Unlike trees they are devoid of view-blocking twigs and leaves and, since they almost always run parallel to roads or railways, they are easily accessible. Migrating birds such as the bee-eaters and swallows use them as gathering areas or as rest sites. Others, such as rollers, shrikes, and the smaller birds of prey, find them ideal lookout posts from which to go on hunting forays. A quick scan of local lines in any new area will normally produce something of interest. I sincerely hope that the spread of satellite communication will not lead to the demise of the humble wire, and that the satellite dish will not replace the TV antenna, another prime perch. Much more technology and the birds will have to perch in trees—ridiculous.

Migration apart, spring sees other things afoot on the natural front. Many of Egypt's larger mammals are producing baby larger mammals. Unfortunately most are elusive, almost by definition. To be an extant Egyptian larger mammal, other than a hominoid, it is an absolute necessity to be elusive. I was made aware of the situation when writing my first article on Egyptian wildlife. I knew it was possible that pitiful remnants of the country's Leopard, Cheetah, Oryx, Ass, and Barbary Sheep populations may remain in some far-flung wadis, and was looking for pictures to illustrate the article. The only photographs I could find were of very dead specimens lying in the desert or draped over hunters' jeeps. The sole cheetah photograph was of a skin from near Siwa. I eventually tracked down two slides of Fennec Fox cubs that had just been dug out of their earth, but someone's knee prominently appeared in each picture. The only other success was a picture of a Long-eared Hedgehog, but a bucket scarcely seemed to be its natural habitat.

However, somewhere in the midst of dense Delta marshes (or what is left of them) Swamp Cats will be having kittens. Those Jackals not yet guests of the Kerdasa taxidermists will be having their puppies in desert-margin dens. In the Eastern Desert, Ibex may be giving birth, while in Sinai the Caracal, a desert lynx, will be having baby caracals. This is probably the most elegant of the larger small cats, a sleek feline, sandy brown in color to render it near invisible in its barren habitat. Its most distinctive feature is its beautifully long and elegant dark ear tufts, far longer than those of its European or American counterparts. Unlike many of its feline relatives, the Caracal is a bird specialist. In Egypt, sandgrouse are probably its main prey. Enormously agile, a Caracal can leap into a flock of sandgrouse killing its chosen victim with a blow of the paw.

Sandgrouse are interesting birds and not in the least bit related to grouse. The resemblance is purely superficial, and even then only with a little imagination. They are relatively stocky, cryptically colored birds with rather pigeon-like heads and long, pointed wings, clearly apparent in flight. Six species are found in Egypt, of which four (Lichtenstein's, Crowned, Spotted and Chestnut-bellied) breed. Curiously for essentially desert dwellers, sandgrouse require water regularly and will fly many miles to drink. Anyone particularly anxious to see a sandgrouse can do far worse than find a remote desert pool, such as the sewage ponds at Sharm al-Shaykh, and watch over it around dawn or dusk. An interesting piece of trivia is that sandgrouse and their closest relatives the pigeons and doves (and perhaps also the extinct Dodo) are the only birds that can swallow. All other birds must, on drinking, raise their heads after each gulp and allow the water to trickle down courtesy of gravity.

***Caracal**. This cat, a rare inhabitant of Egypt's deserts, is hardly ever seen. The long ear tufts show it to be a relative of the Lynx of Eurasia and North America.*

Spotted Sandgrouse *(two males and one female). Sandgrouse have to drink regularly, and often fly long distances to find water. The male in the foreground is soaking his feathers, which will absorb water to be taken back to the young.*

The watery talents of the sandgrouse do not stop here though, for this interesting group of birds has another trick up its sleeve—or rather upper-wing covert. Sandgrouse also breed in the desert some distance from water, and yet the young, like the adults, must drink regularly. To overcome this logistical dilemma the male has a patch of downy belly feathers. On leaving his grouselets in their sandy nest site he rubs his belly against the ground to wipe the feathers clean of preen oil, a substance secreted to keep the feathers in peak condition. On arriving at the waterhole, perhaps as much as 40 kilometers distant, he not only takes a well-earned drink but also soaks his belly in the water. The dry, downy feathers act as a sponge absorbing the precious liquid. The doting father then flies homeward, where the young quench their thirst on the saturated feathers. Prior to hatching, the male uses the water to keep the eggs cool.

In the city, a relative of the sandgrouse, the Palm or Laughing Dove, will be courting. This is the familiar, slimline dove with the blue-gray wings seen throughout Cairo, indeed in virtually every town or village. The male will puff up his neck feathers and strut round his intended, cooing softly. It is this *coo* that supposedly gave them the name Laughing, but to me it sounds nothing like a laugh, more a soothing croon. Air-conditioning units seem to be rather popular strutting grounds for lovelorn Palm Doves. In spring my alarm clock becomes redundant. Every morning I am woken at dawn by the urgent nuptial cooing of an amorous Palm Dove in suitor mode, echoing and reverberating through the hollow metal unit. Alarm clocks, unlike dove hormone, can at least be turned off.

The Palm Dove is not the only bird looking toward raising a family. The Hoopoe is an unmistakable bird, cinnamon pink with black and white barred wings and tail. It has a long, down-curved bill and a crest that is raised in a fan whenever it is excited or when it lands from its moth-like flight. Normally the crest is flattened and, with the bill, gives the head a hammer-like outline. The Hoopoe can be seen jerkily patrolling lawns and gardens throughout Egypt, constantly stabbing at the soil for insects and other prey. Distinctive though it is, scientists do not know quite what to do with the Hoopoe. It is so unlike any other bird that some experts put it in a family all on its own. More commonly it is lumped with the glossy, long-tailed Wood-hoopoes of Africa.

Hoopoes may look attractive but their personal hygiene leaves something to be desired. A pair of amorous Palm Doves may set up home and no-one is any the wiser. Anyone playing host to a pair of Hoopoes is likely to know. The nest smells disgusting. Most birds keep the nest clean by the young birds either instinctively 'going' over the edge or delivering their waste in neat little packages called fecal sacs, which are then removed by the parents. Not so the Hoopoe. Young Hoopoes are reared amongst a steadily rising level of guano that makes their nest quite easy to detect by smell alone. Small wonder that at least two of the world's major religions forbid their adherents to eat the bird.

Spring's big traditional holiday is known in Egypt as *Shamm al-Nisim*. It is a festival with roots beyond either Islam or Christianity, dating back to pharaonic times. The last two *Shamm al-Nisims* I have fled to the country, to

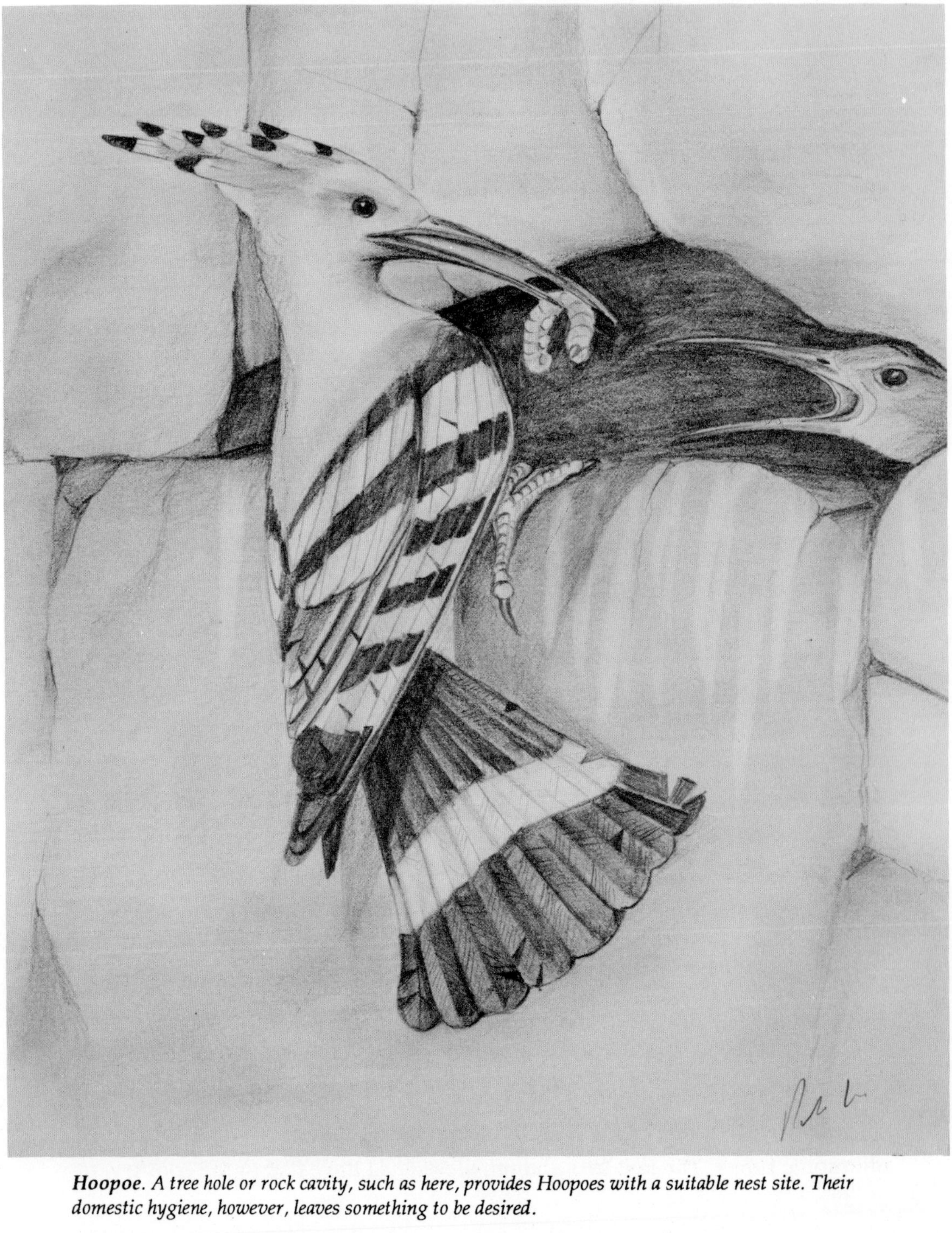

***Hoopoe**. A tree hole or rock cavity, such as here, provides Hoopoes with a suitable nest site. Their domestic hygiene, however, leaves something to be desired.*

a friend's farm to enjoy the traditional foods of egg and onion and endure the traditional dried fish called *fisikh*. The farmlands are still fresh and green in spring. The air is filled with dragonflies and damselflies. Bee-eaters hawk for prey and coucals deliver their melancholy hoots from thick cover.

Two of the most typical birds of the Egyptian countryside both seem full of the joys of spring. The Fan-tailed Warbler is a tiny brown bird, streaked above and with a short, black, white-tipped tail. It is a member of the cisticola group, a large group of confusingly similar warblers found throughout Africa. On the voice front they go in for quantity rather than quality. The trait is reflected in their names. There is the Rattling Cisticola, the Tinkling Cisticola, the Croaking, Wailing, Singing, and even Zitting Cisticola. The Fan-tailed Warbler or Cisticola is the only one of the group to reach Egypt. In habits it somewhat resembles a Skylark, belting out its song from high in the air, the only time it is likely to be seen. The resemblance to a lark stops there. It is hard to imagine even the most stone-deaf poet being the least bit inspired to great verse by the tinny trillings of the Fan-tail. The other bird typical of the farmland is the Graceful Warbler. Like the Fan-tailed Warbler it is diminutive, streaked brown, and prone to stick to thick cover. When it is seen, the long tail normally carried cocked vertically upward is distinctive, giving the appearance of an anorexic wren. In flight the tail is carried in line with the body. Small and dull it may be, but its voice is a give-away, a *trrrr trrrr trrrr* repeated continuously to the point of monotony but occasionally relieved by a lower *zit* being added to each *trrrr*. An insignificantly tiny bundle of streaky gray-brown feathers calling loudly and repetitively from the middle of a low bush is likely to prove to be a Graceful Warbler.

Not all that flies is a bird. Dragonflies and damselflies abound. Both have long, cylindrical bodies. Damselflies are generally more slender and rest with their wings held lengthwise along the body. Dragonflies are usually more thickset and rest with the wings held out at right angles to the body. Some species adopt what is known as the 'obelisk position.' This is not a leaf from some dragonfly *Kama Sutra* but a method of avoiding overheating. The

Fan-tailed Warbler*. I was lucky to find this particular bird perched in a Jointwood tree. Normally they skulk in dense vegetation and are most often seen during the male's display flight.*

Graceful Warbler. *This charming little bird is often seen nervously flitting around in thick cover. The tail, distinctively cocked at rest, is held out horizontally in flight.*

dragonfly will orientate itself so that its back faces the sun. It then raises the long abdomen in the air so that the tip points directly at the sun ensuring that the smallest body area possible is exposed to the full heat. In spring many of these insects are emerging after having spent their formative years as aquatic larvae in irrigation canals, ponds, and lakes. There are many species but one of the most common and easiest to identify is the Scarlet Darter. It is rather small by dragonfly standards but quite chunky in build and bright red.

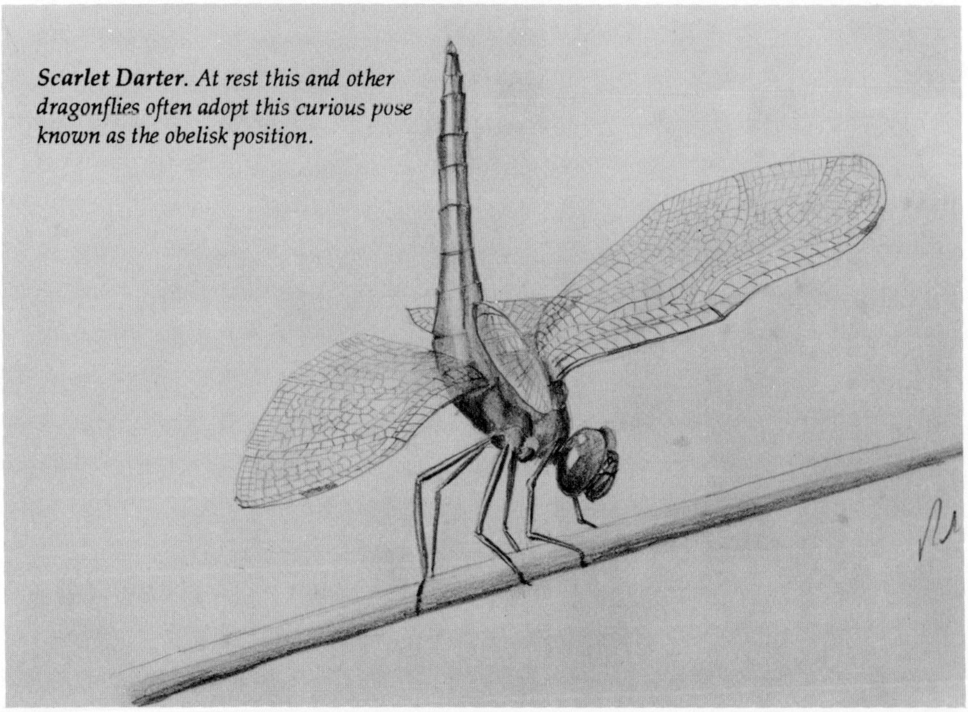

Scarlet Darter. *At rest this and other dragonflies often adopt this curious pose known as the obelisk position.*

As summer approaches, temperatures rise and the reptiles really start to liven up. Some, like the geckoes, will be very evident. The Turkish Gecko is the species most often seen, a familiar house-guest and a very useful one. It is often attracted to outside light fittings because outside lights attract the nocturnal insects upon which the gecko preys. Occasionally they get a bit too attached to outside light fittings. I know of a gecko in Zamalek that took up residence inside the glass shade, a self-defeating move because the light could not then be used, for fear of frying the incumbent reptile. No light, no food. It eventually left. The Egyptian or White-spotted Gecko is an altogether more substantial beast with two pairs of white spots on the shoulders. It is not a city dweller but more a creature of the desert margins. I know of a quite charming pair in Dahshur.

Far less likely to be seen, but equally interesting, are the sand boas. The term 'boa' is likely to conjure up images of huge constrictors in steamy jungles, or of pet snakes flushed down American toilets and growing to vast sizes in New York sewers. Egypt's two species could not be further removed from these mythical serpents. Neither is likely to exceed 80 centimeters in length and both spend much of their time hidden in rodent burrows, often having consumed the rightful occupant. Sand boas are thickset, cylindrical snakes, tapering off at head and tail. When threatened they curl up in a ball, with their head hidden, and raise their tail in the air as a decoy. Unfortunately this rather tame mode of defense has not proved effective against human snake-catchers. Both boas are now strictly protected throughout the year since they play an important role in keeping down rodent populations.

Staying with the less popular members of the Egyptian fauna, scorpions are worth a mention. They are part of a group known as the arachnids, which includes those other falsely maligned arthropods, the spiders. Scorpions are too well known to need detailed description. They have four pairs of legs, the first of which are modified into powerful crab-like pincers, and a long tail that ends in a venomous sting. The venom is generally more unpleasant than lethal (not from experience), but the sting of some, such as the Palestinian Yellow Scorpion of the desert regions, is potentially fatal. Fortunately the venom is injected only in very small quantities.

Scorpions are normally solitary. Encounters with other scorpions often end up with the larger eating the smaller. Life must go on, though, and at some point man scorpion must meet lady scorpion. The affair is more gladiatorial than romantic and in all events brief. However, there is a softer side to the

White-spotted Gecko and Yellow Underwing Moth. *Geckos prey largely on insects. The moth will be hoping that the flash of sulfur-yellow on its hindwings, here just revealed, will distract the gecko long enough to allow it to escape.*

Desert Scorpion. *Scorpions have a poor reputation, due to their painful and sometimes potentially fatal sting. However, the females are unusual among the invertebrates in being good mothers. Here a female carries her new-born young on her back.*

scorpion. The mother shows a degree of maternal care that is almost unparalleled in the invertebrate world. Young scorpions are born one or two at a time and immediately clamber onto the mother's back. The female will then protect and defend them until they pass through their first molt and can defend themselves. Despite their infamy, scorpions are rarely seen. There are plenty around but they are largely nocturnal and secretive, spending their days hidden beneath stones and masonry, in cracks, crevices, and burrows.

Life in the city has some compensations. I arrived home from work one day refreshed and cheerful. Strolling past the back of the Mugamma, that grim monument to bureaucracy and hardly designed to refresh and cheer, I happened to glance up at the sky. The glance was rewarded by a stupendous display of acrobatics as a pair of Kestrels mobbed and harassed a rather stoic Black Kite. The falcons dived and screamed at the larger raptor, who wheeled deftly around, steering with subtle flicks of a long forked tail. My heart lifted—

a thrilling dogfight between two species of birds of prey in the center of one of the world's largest and most densely populated cities. Sometimes though, one has to escape.

The wonderful thing about the natural world is that it is so full of surprises. However much one reads, observes, and records, the unexpected, the unusual, or the plain bizarre can happen. One weekend I was down in Wadi al-Rayan, south of the Fayoum. Wadi al-Rayan is an unusual place, flooded with overflow water, and there are now two large freshwater lakes in the midst of the desert connected by a narrow channel. The second lake is somewhat lower than the first, resulting in a waterfall some five meters high. Five meters is hardly a Niagara but this, with the exception of ephemeral desert streams and the turbine outlets at the Aswan High Dam, is Egypt's highest waterfall. It is surrounded by dense reedbeds full of Clamorous Reed Warblers being clamorous, Graceful Warblers being graceful, and Yellow Wagtails being—well, yellow and wagging their tails. On the lower lake rafts of Tufted Ducks and Coots in Puritan black and white bobbed alongside Great Crested Grebes just coming into their flamboyantly Cavalier nuptial plumage. Along the shore a solitary Squacco Heron in buff and white stole furtively. You don't often see them in the open.

An amble along the lakeside revealed more ducks—Widgeons, Pochards and the elegant Pintail. It was idyllic. The lakewater gently lapped against the shore. Underfoot, the gravel crunched and a small stone ran across a nearby sand dune. Something was clearly amiss. Small stones should not run across sand, or anything else for that matter. Granted, given a certain roundness and an adequate slope, one might expect a roll. But this was level sand and the stone looked flat. Under these conditions any stone worth its salt should be stationary. Another step and the stone once more sped off, this time with me in pursuit. I eventually cornered and caught the stone and found in my hands not a particularly fidgety lump of rock but an insect (three pairs of legs) the like of which I had never seen before. It was pinkish-gray with a short, very flattened body with a clearly segmented abdomen and vestigial wings. The eyes

Squacco Heron. *Though the Squacco Heron is normally a skulking bird of dense reedbeds, I was lucky enough to see this individual out in the open at Wadi Rayyan.*

were large and the same color as the body. The back two pairs of legs were fairly long and powerful as befitted a scuttler, but the front pair were bent in the manner of a praying mantis. Indeed, had a praying mantis the misfortune to be stamped on, I believe the result would be fairly similar to what I had in my hand.

Subsequent research has shown my 'stone' was indeed a praying mantis, not squashed but as nature intended. It was a desert mantis bearing no common English name but known, doubtless in exceedingly limited circles, by the Latin name *Eremiaphila.* Though reportedly very rare in collections, specimens had been found by the French naturalist Lefebure between the Nile and Bahariya in the 1830s. These desert mantises are considered to be more

***Tufted Ducks and Great Crested Grebe.** Grebes can be distinguished from the superficially similar ducks by their sharp bills and the way they generally lie lower in the water. Both the Great Crested Grebe (back) and the Tufted Ducks (two males and one female) here are winter visitors and will soon be off to northern Europe.*

highly adapted to desert life than virtually any other animal. They are able, according to the Cambridge Natural History, written in the heady prose style of nineteenth-century natural historians, to live "in places where no vegetation exists and to assimilate in appearance with the sandy soil so that individuals agree in tint with the soil on which they dwell." This one was doing just what the book said. I was chuffed. I have found little else about *Eremiaphila*, so any further reports would be gratefully received.

Through spring the migration continues, with a wide variety of birds flapping, gliding, and soaring their way north to European nesting grounds. Other birds are beginning to turn their thoughts to courtship here. High above the city center wheel screaming flocks of Pallid Swifts. As spring progresses the flocks get distinctly larger and noisier. For swifts courtship is an aerial affair conducted, like everything else in a swift's life, at high speed and altitude. They will soon be nesting in rooftop crevices throughout Cairo but are comparatively local elsewhere.

Another bird worth looking out for at this time of year is the Blackbird. Over most of Europe the Blackbird is treated by birdwatchers with a familiarity that falls well short of contempt but borders on apathy. The female is pretty dowdy, but the male cuts a bit of a dash in pure black with a bright yellow bill. In Egypt they are restricted as a breeding species to the Delta and northeastern Sinai. However, they are thought to be expanding their range south. I have seen them on several occasions at Abu Sir and Saqqara but have been unable to prove nesting. Each spring I'll be keeping my eyes peeled and my ears pricked. The male may be dressed in somber black but his fluting song, poured out to announce his territory, is anything but funereal. Happily this little piece of melody against the city's cacophonic backdrop is becoming increasingly common.

As spring draws on, the magnificent flypast of birds great and small begins to peter off. However, even toward the end of May late migrants, including some of the larger eagles, can still be seen passing through. However, Nature is anything but predictable. One can leave the armchair, root out the binoculars, pick up a field guide, and head for somewhere seemingly irresistible to anything on wings—and still end up seeing absolutely nothing. One can read here, or elsewhere, of the thousands of raptors streaming through Egyptian airspace and end up staring at skies cloudless, blue, and totally devoid of anything feathered. The vagaries of weather and thermal conditions, crucial to the larger, broad-winged migrants, offer a partial explanation. A dearth or abundance of food in the wintering areas may advance or delay a bird's departure. But the entire blame for birdless skies cannot be attributed to climate and food. Thousands of birds passing overhead on outspread wings of up to three meters can be very hard to see. Many of the birds of prey fly very high and silently. While other species such as the Common Crane have the courtesy to trumpet their passage with a clear, repeated *krook krook*, the raptors soar past with sealed bills and often at such altitude that they appear to the naked eye as little more than specks. It is a question of finding in the wide blue yonder that first speck. Once that is found, and preferably magnified with the aid of binoculars, others become apparent, and then more until the sky is full of specks and you wonder how on earth you missed them in the first place. The binoculars become helpful, indeed vital, in differentiating one high-flying speck from another. Raptors are notoriously difficult to identify.

There is still every possibility of seeing nothing, however, so it is probably wise to head somewhere like 'Ain Sukhna on the Red Sea. Here, even if the birds do not show up, there are the coral reefs and, for the observant, the chance of seeing small schools of Bottle-nosed Dolphins making their leisurely way along the coast.

By early June, most of the birds seen are likely to be staying, and, as the weather warms, thoughts of humans and animals alike turn to surviving another sweltering Egyptian summer.

Summer

It hardly takes the foresight of a clairvoyant to predict that the Egyptian summer will be hot. Come early June the TV weather forecasters slip into a four-month overdrive, announcing each evening in hushed tones to an unsurprised audience that "Tomorrow will be hot in the north and very hot in Upper Egypt." Anything to the contrary would have been in the news headlines a few minutes earlier. Cairo will be hot. Other areas, particularly in the Western Desert and to the south, will be horribly, oppressively hot. Humans have a variety of ways of keeping cool, ranging from cold drinks to fans and culminating in those noisy metal boxes known as air-conditioners. Animals too must cope with the high temperatures, particularly those of the desert. Like a machine an animal's body must not overheat, so the first problem is how to remain cool. The second, related problem is that of dehydration. Sweating, a natural cooling mechanism all too familiar to humans, is not an option open to many desert species since it involves the loss of too much precious moisture.

Egypt's deserts are populated by a whole host of small mammals. Representatives include the aptly named Charming Dipodil, the Large North African Dipodil, the Silky Jird, and Henley's Gerbil. All these mouse-like creatures have adapted to living in desert environments where daytime temperatures can exceed 50°C, the only food is dry seeds and fruits, and there is no standing water.

The most straightforward way to keep cool is simply not to go about during the heat of the day. Most desert creatures are nocturnal, the huge eyes of pet gerbils (originally from Mongolia), for example, being an adaptation to true night life. The day is spent in burrows where the temperature can be as much as twenty degrees lower than that above ground. Not only are the burrows cooler during the day but, despite extreme fluctuations in temperature in the desert between night and day, the temperature in the burrow remains extremely stable.

Water is the desert's most precious commodity, and its rarest. Many of the small desert mammals obtain sufficient moisture from their food. Some adhere to a diet not normally associated with rodents, living on insects, small reptiles, and even other small rodents. Yet even those mammals, such as the Lesser Gerbil, that feed exclusively on dry vegetation manage to glean enough moisture from their food to survive without ever actually having to drink. One way to obtain moisture is to feed at dawn when the food plants are impregnated with dew. However, such minute traces of water would be insufficient were it not for adaptations in the animal's metabolism. Avoiding the heat of the day in their subterranean dens, gerbils, jirds, jerboas, and other such small mammals do not need to perspire. Their kidneys function far more efficiently than our human kidneys do, producing only a few drops of much more highly concentrated urine. Furthermore, the burrows are not only cool but humid. By taking their dew-soaked seeds back to their burrows the animals minimize the moisture lost in evaporation. All terribly clever but, unfortunately for the jirds, gerbils, and dipodils of this world, wherever there are small animals there are bigger ones to eat them.

Since most of their prey species are nocturnal, most of the desert predators are likewise so. An active night life might solve problems generated by high daytime temperatures, but it does not spare the foraging dipodil the attentions of the fox, jackal, or wild cat. Of these the Fennec Fox is the most highly adapted to desert life. The world's smallest, and arguably most attractive, vulpine, this diminutive predator hunts by night, finding its food by sight and sound. Fennecs are around 60 centimeters in length, of which over a third is a bushy tail, and weigh about one kilogram. They have typically foxy faces and

Fennec Fox. *The world's smallest fox, the Fennec seems to be all ears and eyes—adaptations to its nocturnal habits.*

a beautiful, thick creamy-white fur. As befits a nocturnal predator, the brown eyes are large. If its eyes are big, then its ears are disproportionately enormous, but hearing is only one of their functions: they also serve as natural radiators through which the fox can loose heat to the air. Fennecs do not need to drink. They obtain all the moisture they need from the body fluids of their prey. Fennec Foxes are uncommon but widely distributed, especially in the Eastern Desert and Sinai.

The Fennec Fox is an active hunter. Other predators rely on ambush in order to catch their prey. The Egyptian deserts are home to the Horned Viper, a thickset snake with a broad head, that reaches some 60 centimeters in length. It is generally colored in muted shades of beiges, yellows, and creams to match its habitat. It gets its name from the modified scale that forms a horn over each eye. Horned Vipers feed on small reptiles and young rodents. The snake has the ability to literally sink into the sand until it is completely hidden but for the eyes. The horns are not decorative but serve to keep the eyes free

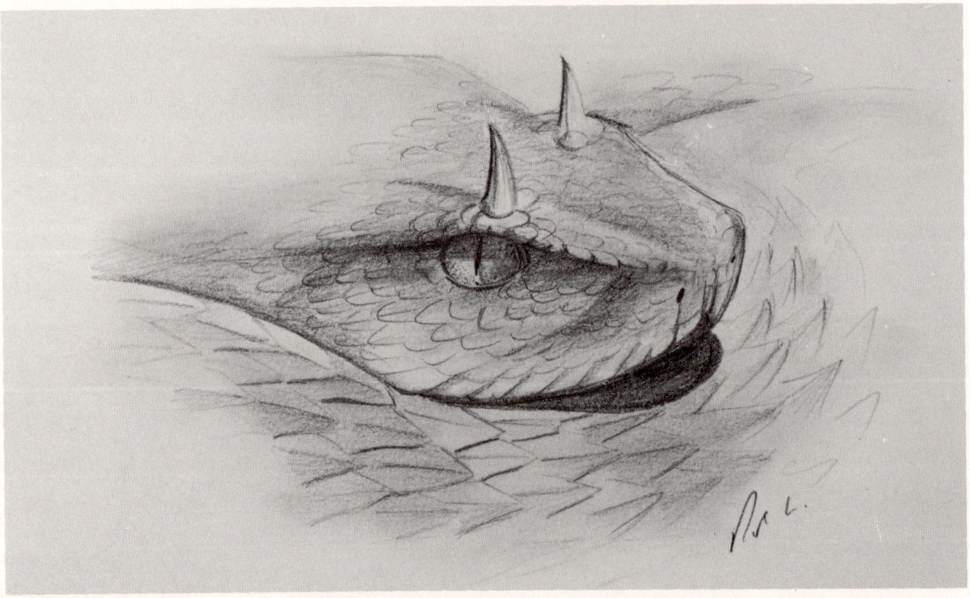

***Horned Viper**. The Horned Viper is a desert dweller. The modified scales over the eyes that give it its name are functional rather than ornamental.*

of sand once buried. The snake will then lie all but invisible until some hapless creature comes too near and the viper lunges out, killing its victim with venom delivered in a lightning strike of its fangs. Horned Vipers will bury themselves not only to catch prey but also to keep cool and as a defense measure.

A curious tale relating to this species comes from the Victorian naturalist John Anderson. He reported that these snakes were eaten in certain areas of Cairo—but live and tail first. His description makes grim reading. Apparently the consumers suffered no ill effects, but it is scarcely to be recommended.

Desert rodents are also in danger from the air. Two species of owl frequent the Egyptian deserts, one widespread but uncommon, the other extremely local. The former is the Eagle Owl. The Eagle Owl is a huge bird, around 70 centimeters long, and size alone should distinguish it from any other owl species. There are two forms in Egypt, a darker one of the desert margins and a pale desert form. In both the eyes are huge, about the size of a human's, orange and staring. The ear tufts are long and prominent. Owls hunt by sound more than by sight, but although many species possess ear tufts, these are merely elongated feathers around the ear region. They have nothing to do with hearing. Eagle Owls inhabit the desert regions throughout Egypt but are strictly nocturnal and rarely seen. By early summer they will be rearing their owlets. Owls in general are economical on the home front. Not for them an elaborately woven nest of cobweb and plant fiber. A scrape on the ground, a sheltered crag, or the deserted nest of some other bird suffices. The eggs are not egg-shaped but spherical and invariably plain white. The number of young reared depends on food availability. Owl eggs are laid at intervals of several days and hatch over a corresponding time period. Thus an owl's nest, should it be elaborate enough to warrant the term, may contain several owlets of different ages and sizes. In a good year, with plenty of food, all the young may survive. In a poor year the smaller, later-hatching young will be unable to compete for food with their larger, older siblings and will die.

Desert Eagle Owls nest in rock crevices or small caves. There were reports earlier this century of a pair that nested on the Great Pyramid. Two nightly

Eagle Owl. *This is one of the world's largest owls. Despite its size it is difficult to observe in the field due to its cryptic coloring and nocturnal habits. This bird, which I watched for over an hour in the early morning, did nothing more energetic than blink.*

showings of the Sound and Light have probably forced them elsewhere. It may not have been a good year for Eagle Owls in 1992. A pair I observed were rearing a single owlet, a great fluffy thing that looked more like a sheep with a hooked bill than a young bird.

The other desert owl is Hume's Tawny Owl, one of the least known owls in the world. Apart from a single record from the Eastern Desert it is restricted to southeastern Sinai. It has been reported calling around the Monastery of St. Catherine, a steady *hoo, ho-hoo, hoo, hoo.* Hume's Tawny Owl has the rare distinction of being one of the very few birds not to have been shot by Richard Meinertzhagen. Meinertzhagen was the author of the seminal work on Egyptian birds, *Nicoll's Birds of Egypt*, published in 1930. As was the norm in those days of gunsight natural history, most of the work was done from collected specimens. You had not seen something unless you had shot it and had it in your hand. A mere sight record meant little. Meinertzhagen was camping in Wadi Firan in the Sinai in 1928 when he saw a Hume's Tawny Owl perched on a palm. To be seen by the great naturalist normally marked the death knell of anything feathered, but not on this occasion. As the author himself wrote: "The desire to observe such a rare bird in its native haunts was accomplished at the expense of a much-coveted specimen, for after five minutes the Owl decided to move on." It must be a truly remarkable bird.

Apart from the owls there is another group of birds active after dark, the nightjars. Egypt boasts three and a half species. The European Nightjar is a rare migrant, the Nubian Nightjar has been found only in the southeasternmost corner of the country, and the Egyptian Nightjar is an uncommon breeder. The half is the Red-necked Nightjar. In the early nineteen-eighties one reportedly crashed into a ship in Port Said, but the experts find the record unacceptable. Though the Egyptian Nightjar supposedly breeds on the flat rooftops in the City of the Dead, I have never found it there. For me the ultimate nightjarring spot is Abu Sir, north of Saqqara. The region around Abu Sir is best explored on horseback. Not all horses are great naturalists, generally being more interested either in getting home or in other horses, but the odd one shows a

Egyptian Nightjar. *This rather elusive bird spends the day virtually invisible lying in the sand. At dusk they take to the air and hawk for insects, flitting around like giant moths.*

real keenness. I refer particularly to Masaoud. It is to Masaoud that I owe my best views of Egyptian Nightjars. Nightjars are slim birds with long wings and tail, like slender hawks—indeed their occidental relatives are known as nighthawks—and are characterized by a tiny bill and huge gape. The Egyptian Nightjar is colored in cryptic beiges and sandy browns, relieved by a white throat. By night they hawk ghostlike for insects, by day they sit invisible on areas of sandy ground, relying on the extreme effectiveness of their camouflage. Not surprisingly they are rarely seen. Masaoud is a glossy bay horse with a very long mane and the uncanny knack of finding these birds. I have seen Egyptian Nightjars three times now. The first time it was a single bird at a distance. On the other two occasions flocks flew up from beneath the hooves of Masaoud. So confident are they of their camouflage that they will sit tight and invisible until virtually trodden on. It is this ability to almost tread on nightjars that has rendered Masaoud so useful. On both occasions I have looked round in sheer delight as the air filled with flushed nightjars wheeling and circling before settling, once more invisible, against the desert sand.

Night life in the desert seems pretty lively. However, the abundance of life in the desert is never apparent by day since most creatures are resting, roosting, or slumbering below ground. For the human visitor to these arid areas, waking up after a night in the desert can prove an interesting experience. Any tracts of sand are literally riddled with the tracks of those creatures active at night. With some experience it can be possible to reconstruct the little dramas that were being played out in the darkness. On one occasion, for instance, I came across parallel lines of apparent pinpricks in the sand that finished abruptly as they met a coarse double furrow carved in the dune. The furrows, also parallel but larger than the line of pinpricks, looked as though some attempt had been made to sweep them smooth. From this intersection the furrows continued, but shallower and neater and with distinct pits indicating footprints. The likely scenario was of a desert beetle out on a nocturnal forage (the pinpricks) being discovered by a small lizard. The lizard ran at the beetle, hence the rather coarse furrows, its tail swishing from side to side flattening the furrows out somewhat. The consumption of the beetle explains the disappearance of the pinpricks. The lizard then continued on its way in a more leisurely fashion, leaving a less messy set of tracks. All this inscribed ephemerally upon the sand.

One of the few desert creatures that can be seen during the day is the Fat Sand Rat, an extremely unendearing name for a rather endearing creature. Fat Sand Rats are like overgrown gerbils, though with less disproportionately large hind limbs and a moderately long black-tipped tail. I have watched sand rats in Wadi Digla, just outside Cairo, going about their business in broad daylight. Their business seems to be gathering food and taking it back home for more leisurely consumption. For the sand rat, home is a burrow, often at the base of a small shrub or tamarisk and characterized by a mound of sand and droppings. From here it will emerge cautiously, taking a good look round, before scuttling out to collect some juicy piece of vegetation and scuttling back.

Owls and nightjars apart, the birds are one group of animals not particularly well represented in the desert proper. There are few species of truly desert bird. Few birds burrow, and so they are generally unable to avoid the heat of the day. While their body temperature is naturally higher than most mam-

Fat Sand Rat*. The seemingly lifeless desert is full of small rodents, though the Fat Sand Rat is one of very few active by day. This one was seen at Wadi Digla.*

mals, so that they have less need of cooling (they loose heat through panting), birds are more limited by food and water requirements. Sandgrouse, for instance, fly many kilometers each day to congregate at waterholes. Others, such as Desert, Bar-tailed Desert and Hoopoe Larks, somehow manage to survive on moisture obtained from their diet of apparently dry seeds.

Of these desert denizens, it is the Hoopoe Lark that is most likely to catch the eye and, more especially, the ear. The Hoopoe Lark is an aptly named bird, clearly a lark but with a rather long, slightly down-curved bill and strikingly black and white wings that give it a somewhat Hoopoe-like appearance. It is also rather common. Amongst naturalists, familiarity often leads to disinterest. The sparrows and mice and other ubiquitous beasts are all too frequently ignored in quests for rarer game, but in doing so much can be missed. I had seen Hoopoe Larks. They were ticked off on my checklist, noted in my diary, and filed in my memory. At one stage they had even warranted the odd sketch.

If anything of that type appeared as I scanned the desert wastes with my binoculars I would barely pause. I had, I thought, done the Hoopoe Lark. Then news came from Saqqara of the 'Dive-bomb Bird,' a curious winged beast that would drop straight from the sky while emitting a high pitched whistle.

It was some weeks before I saw the 'Dive-bomb Bird' for myself. In the desert just beyond Abu Sir my attention was alerted by a curious musical piping and I turned to see a small shape fall hurtling toward the ground. Before long, from the same patch, what was now clearly a bird rose from the ground in a spiraling climb, piping furiously only to peak, fold its wings and plummet once more to earth. The wings, strikingly black and white in contrast to the pale beige body, were diagnostic. The bird was a Hoopoe Lark and what I had witnessed was the male's territorial display flight. Clearly I had not 'done' the Hoopoe Lark. Early June seems to be the best time to catch this rather spectacular performance from a seemingly unspectacular little bird, but he performs matinees through July.

***Hoopoe Lark**. The Hoopoe Lark is probably the least lark-like of the larks. It is a common bird around Saqqara, drawing attention to itself through its spectacular display flight.*

Those desert birds that do manage to eke out a living in their less than lush habitat are generally smaller and paler than their counterparts from more fertile pastures. The Desert Lark is a small finch-like bird, a lark (not surprisingly) found in deserts. It is soberly colored in uniform beige, but the shade varies between populations, changing to match the habitat (in the lava desert of northern Jordan there is a population that is almost black). Eagle and Little owls too demonstrate this phenomenon, some desert forms being strikingly pale.

Not all desert birds, though, are muted and beige. The odd ones show splashes of color. The Trumpeter Finch, named for its call, and the Sinai Rosefinch are both tinged with pink. One bird, the White-crowned Black Wheatear, is pitch-black with a patch of white confined to the crown. Why it should be this color is unclear. However, a White-crowned Black Wheatear perched against a rocky cliff backdrop looks just like an area of deep shadow. Then there is the Sand Partridge. This small gamebird ekes out a living amongst the rocky desert screes. The male is tinged with lovely muted shades of purple, pink, and maroon. The female, it has to be admitted, is more uniformly sandy, relieved by an orange bill and dark eye.

Pale coloration has a clear advantage in the desert from the camouflage stance, but it is also cooler. Anyone who has sat through a Cairo traffic jam in midsummer in a black taxi is aware of the heat-absorbing qualities of the color black. It takes very little time for the vehicle to attain the temperature and comfort of an overworked blast furnace. It is therefore somewhat curious that two of the largest desert birds, the Brown-necked Raven and the Fan-tailed Raven, are both sizable and pitch-black. The former is the archetypal crow, and more crow- than raven-sized, all black except for a brown neck—or so the name leads us to believe. In fact the brown neck is rarely visible in the field and even in museum specimens requires a little imagination. Any large black bird in the desert is likely to be a Brown-necked Raven. The Fan-tailed Raven is a more aptly named bird, all black but with a very distinctive, very short tail, giving it the flight silhouette of a ghoulishly large bat. It is only likely to be seen in Sinai. As they are both aggressive, scavenging, and predatory birds, they would seem to gain little advantage from cryptic coloring, but they should be grossly overheated. The key lies in the feather. A remarkable thing the feather. Engineers have yet to develop a structure quite so light that is quite so strong. However, it is also a superb insulator. Enterprising zoologists have demonstrated that although the temperature of the outer feathers of a Fan-tailed Raven is indeed very high, their insulating capacity is such that skin-deep the birds are comparatively cool.

While on the subject of ravens and crows, the familiar urban crow, the ubiquitous black and gray creature with the guttural *caaww*, is the Hooded Crow. The model opportunist, it is confined to the Delta and Nile Valley, including the centers of Egypt's largest cities. While many birds shun human company, the Hooded Crow actively seeks it, or rather the garbage and refuse that the human presence necessarily implies. Brash, loud, and bold it is doing very nicely thank you.

While nature has evolved a number of clever ploys to keep cool, some creatures seem to bypass all this. If humans have air-conditioned apartments then they simply move in with us. Our artificially maintained dwellings can duplicate an animal's natural habitat. Thus one evening I welcomed a cricket into my flat. It was a small, glossy black insect with long, shiny hind legs and equally long, slender antennae. It was going through some form of toilet, stretching its ample hind limbs against its abdomen as if to make them even glossier. The legs gave it away. It was a cricket and, feeling that the others in the room might like to know of its presence, I announced it as such. After introductions it promptly took off, to cries from a skeptical audience of "But crickets don't fly!"

As problems go it is a bit of a knotty one. Most crickets have wings but in many species the hindwings are reduced to such an extent that they are either absent or useless for flight. Even species that can fly can do so only when adult. You can, should you feel so inclined, age a cricket by its wing development, young ones being called instars.

However, crickets are renowned not for their powers of flight but for their singing. Unlike birds, which warble pleasantly in the conventional manner via throat and mouth, crickets stridulate. A cricket's 'song' (and it is only the males that 'sing') is produced by rubbing the comb-like underside of the right forewing over the edge of the left forewing. If it seems somewhat odd to sing with the wings, it is perhaps odder that crickets hear through their legs. The ear is a small membrane on each of the large back legs. Anyone unfamiliar with the song of the cricket should stand on the platform of Sadat Metro Station. The air is filled with the urgent chirps of bachelor crickets. These will probably be Black Crickets. Presumably the darkness of the tunnels is particularly conducive to cricket nuptials. By May, the hot, dust-laden winds known as the *khamasin* will have got them started and they can be heard singing through to November and early December. The other common cricket is the House Cricket, abundant in and around buildings, particularly in agricultural areas.

If familiarity threatens to dull one's enthusiasm for the rather mundane, it can do the same for the relatively exotic. As summer progresses the waters of the Red Sea become increasingly attractive for human and beast alike. Around late May and through June vast numbers of seabirds will be gathering to breed on the Red Sea islands off Hurghada. For many of these birds, the arms of the Red Sea represent the northernmost and westernmost limits of their ranges. To them, the warm waters of the Red Sea are merely an extension of the Indian Ocean. For birdwatchers, particularly those from northern climes where taking your coat off on a beach is a sign that summer has arrived, the Red Sea offers a selection of avian exotica.

For those who think that a seagull is a seagull is a seagull, the Red Sea islands prove something of an eye-opener. The two species of gull that breed here are quite unlike the pallid white and gray gull stereotypes with yellow bills that screech and scream over quaint little northern fishing harbors. The Sooty Gull lives up to its name. It is predominantly brown, darker on the head and mantle, and with a dirty white belly and tail. The White-eyed Gull is similar, but is a rather more dapper bird with a paler neck and breast and a darker and

White-eyed Gull. *A real Red Sea speciality, the White-eyed Gull is more common here than the similar Sooty Gull.*

more distinct hood. The bold white patch over the eye gives it its name. White-eyed Gulls are locally the more abundant and can even be seen around the main tourist haunts. A good place to meet up with them is on Giftun Island. The boat ride out to the island should yield the odd White-eye gliding on narrow, rather angular wings and being made to look a little clumsy by flocks of slightly more graceful Lesser Crested Terns. On arriving on Giftun most visitors are ushered to the ethnic barbecue shack of their respective boat. Resist the food for a minute or so and wander round the back of the shacks. The vista is unattractive, but then it's not one that is meant to be seen. The accumulated non-biodegradable refuse from every other barbecue is heaped in piles. The only advantage of these unsightly heaps is that they seem popular scavenging grounds for the local White-eyed Gulls. Very prepossessing birds in an unprepossessing landscape. This relative local abundance of the gulls should be put in perspective. The White-eyed Gull is virtually endemic to the Red Sea, and is not found elsewhere, so the island colonies are of international importance. Return home with tales of White-eyed Gulls and friends in the know should be impressed. With this in mind, on a recent jaunt, while scanning the waves for other local specialties such as the Brown Booby and the Red-billed Tropicbird and seeing only yet more White-eyed Gulls, I contented myself with the thought that, although they were common there, I could see them nowhere else.

The Brown Booby and the Red-billed Tropicbird are both near relatives of the pelican but look like neither the pelican nor each other. The Brown Booby is a chocolate-colored gannet. It owes its name to the fact that in the past it was so tame that hungry sailors could just stride ashore to the breeding grounds and club the nesting adults to death. The Red-billed Tropicbird owes its name to the fact that it has a red bill and lives in the tropics. It is a large white bird, barred with black and with a large crimson bill. Its most distinctive feature, though, is its slender, white tail-streamers that make up over half of its length. Essentially oceanic, tropicbirds are only likely to be seen close to land when they are breeding.

***Brown Booby**. The rather comic name of the Brown Booby seems appropriate on land, with its rubber-footed clumsiness. It is, however, an adept flier and skillful fisherman.*

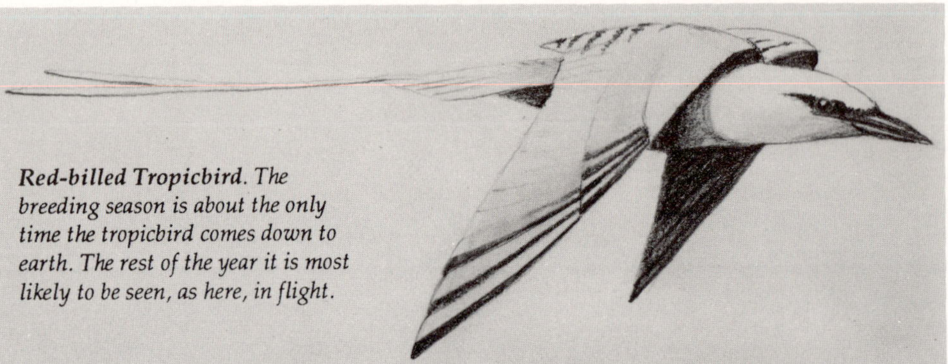

Red-billed Tropicbird. The breeding season is about the only time the tropicbird comes down to earth. The rest of the year it is most likely to be seen, as here, in flight.

The booby and the tropicbird both eluded me on my last trip, as did the dolphins. Dolphin sightings, principally of the Bottle-nosed Dolphin, are supposedly common on the boat trips out to the islands. In keeping with the theory that the more interested the observer the less likely an animal is to reveal itself, they failed to show up. I homed in on every break in the waves in the vain hope that it was a dolphin's dorsal fin breaking the surface, but to no avail. I sought to outwit them. I feigned total disinterest to lure them forth, but not a flipper was to be found. I have had better luck with dolphins further north around 'Ain Sukhna, and also off Sinai. The headland at Ras Muhammad has proved pretty productive on the dolphin front. Small schools can regularly be seen, though swimming some way off the reef out in deeper water. Again, the species seen is generally the Bottle-nosed Dolphin, familiar through the now discredited dolphinariums and through such revoltingly anthropomorphic TV series as *Flipper*. See wild dolphins swimming leisurely along the Red Sea coasts and you'll never want to see one leaping through a plastic hoop in a glitzed-up swimming pool again.

Underwater, things were a little brighter. All the inshore reefs around Hurghada have been systematically destroyed by hotel developers building hotels to house all the people coming to Hurghada to see the reefs. If the logic of this escapes you, join the club. However, out around the islands off Hurghada there are still rich coral communities, though even these are under increasing pressure from tourism. On the plains of Africa, the trained observer can find a lion-kill by homing in on the circling vultures. Off Hurghada the trained snorkeler can find a rich area of coral by homing in on the circling dive-boats. For snorkelers, the location of a large moray or a friendly Napoleon Fish can best be pinpointed from above water. Simply look to where all the little fluorescent snorkel tubes are congregating and there is bound to be something of interest, or else someone has lost a flipper.

Bottle-nosed Dolphin. *This pair was part of a small school seen off 'Ain Sukhna.*

Despite all this the reefs are stupendous. Off Little Giftun island, parrotfish were grazing like herds of gaudy, submarine sheep. There is a basic parrotfish body plan. They are stout, bullet-shaped fish with a rounded snout bearing prominent teeth. The tail is either square or lunate (longer at the edges than in the middle). But within this very basic blueprint it appears that, despite their numbers, no two parrotfish are the same. Firstly, there are a number of different species. There are Rusty Parrotfish, Steepheaded Parrotfish, Bridled, Palenose, Bullethead and Bluebarred Parrotfish, to name but a few. Matters are then somewhat complicated by the fact that young, male, and female parrotfish of all species are differently colored and patterned. While the female Bluebarred Parrotfish is a greenish brown, barred laterally and (predictably) in blue, the male is a fairly uniform pale aquamarine with neither hide nor hair—or scale—of a blue bar.

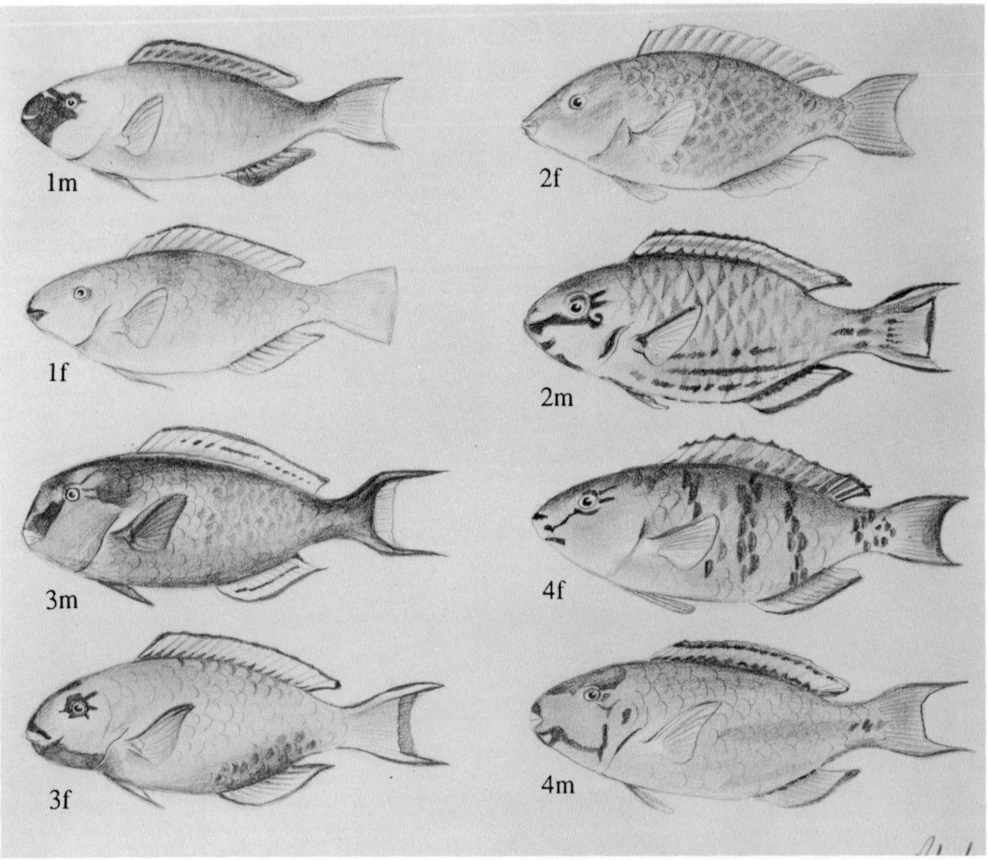

Parrotfish. *These four species show something of the bewildering variety of the parrotfish. Males and females are shown here, but changing sex (and hence pattern) is one of the group's characteristics.* **1**. *Rusty Parrotfish.* **2**. *Palenose Parrotfish.* **3**. *Steepheaded Parrotfish.* **4**. *Bluebarred Parrotfish*

Matters are then totally complicated by the fact that a parrotfish does not always stick to the sex it was born with. Humans, and indeed most of the higher animals, are born one of the sexes and continue in that form throughout their natural span. The parrotfish, along with its relatives the wrasses, scorns such conventionality. All parrotfish are born female and then, as time passes, they gradually become male. With certain wrasses, and presumably with the parrotfish, this is territorial. A dominant male wrasse will hold a territory and defend it vigorously against other male wrasses. In that territory he will have a harem of females. The pressures of hanging on to lady wrasses and territory will ultimately prove too much, and the male will at some point die. On his demise the dominant female wrasse will change sex and become male, and take over the territory. As far as identification goes, it makes wrasses and parrotfish very tricky, particularly when you add the complication of fish in transition, halfway through changing from female to male.

The parrotfish's name derives from their gaudy colors and their sharp mandibles that bear an uncanny resemblance to a parrot's bill. With these they grind away at the stony coral to get at the algae associated with the soft

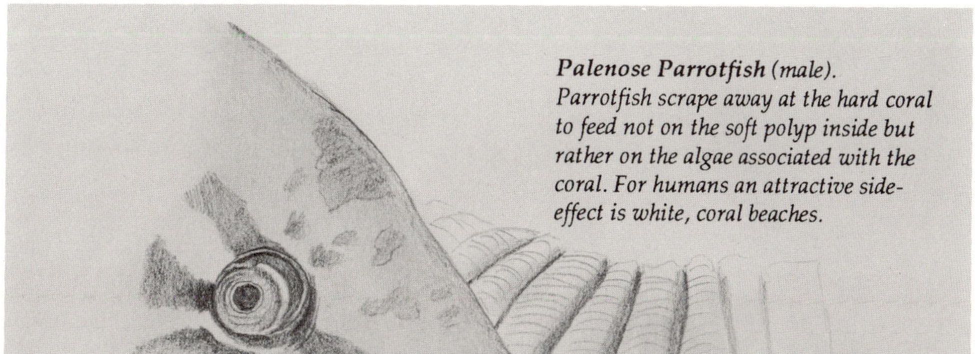

***Palenose Parrotfish** (male).*
Parrotfish scrape away at the hard coral to feed not on the soft polyp inside but rather on the algae associated with the coral. For humans an attractive side-effect is white, coral beaches.

animal living within. Even the most ecologically disinterested beach-bum owes a great deal to the parrotfish. Much of the tropical white sand that graces coral beaches worldwide has passed through the gut of a parrotfish. They consume the algae and excrete the inedible hard casing ground down into grains. These grains, washed ashore, help form the swathes of white sand so beloved of travel brochures and heliophiles. While most reef residents are silent, parrotfish can be heard as they scrape away at the coral.

One of the more striking fishes of the Red Sea, and readily seen around Little Giftun, is the Picasso Fish. Its curious name comes from its bizarre color scheme, yellow-green abstractly marked with blue, black, and red stripes. The Picasso Fish is one of the triggerfish, a group so named from the erectile spine on the dorsal fin. When in danger the fish rushes for a reef crevice, squeezes in, and erects the 'trigger,' making it nigh on impossible to remove. The Picasso Fish's closest relative is a Hawaiian triggerfish known to the Hawaiian islanders as the Humuhumunukunuku-a-puaa. The name has, oddly, never caught on here.

The final reef denizen worthy of special note off Giftun is the Giant Moray. Any creature serpentine in form seems destined to be maligned. No doubt if

Picasso Fish. *The remarkable patterning that gives this fish its name serves to break up its outline against the coral and hence protects it from predators. If this fails it has a second line of defense in its 'trigger.'*

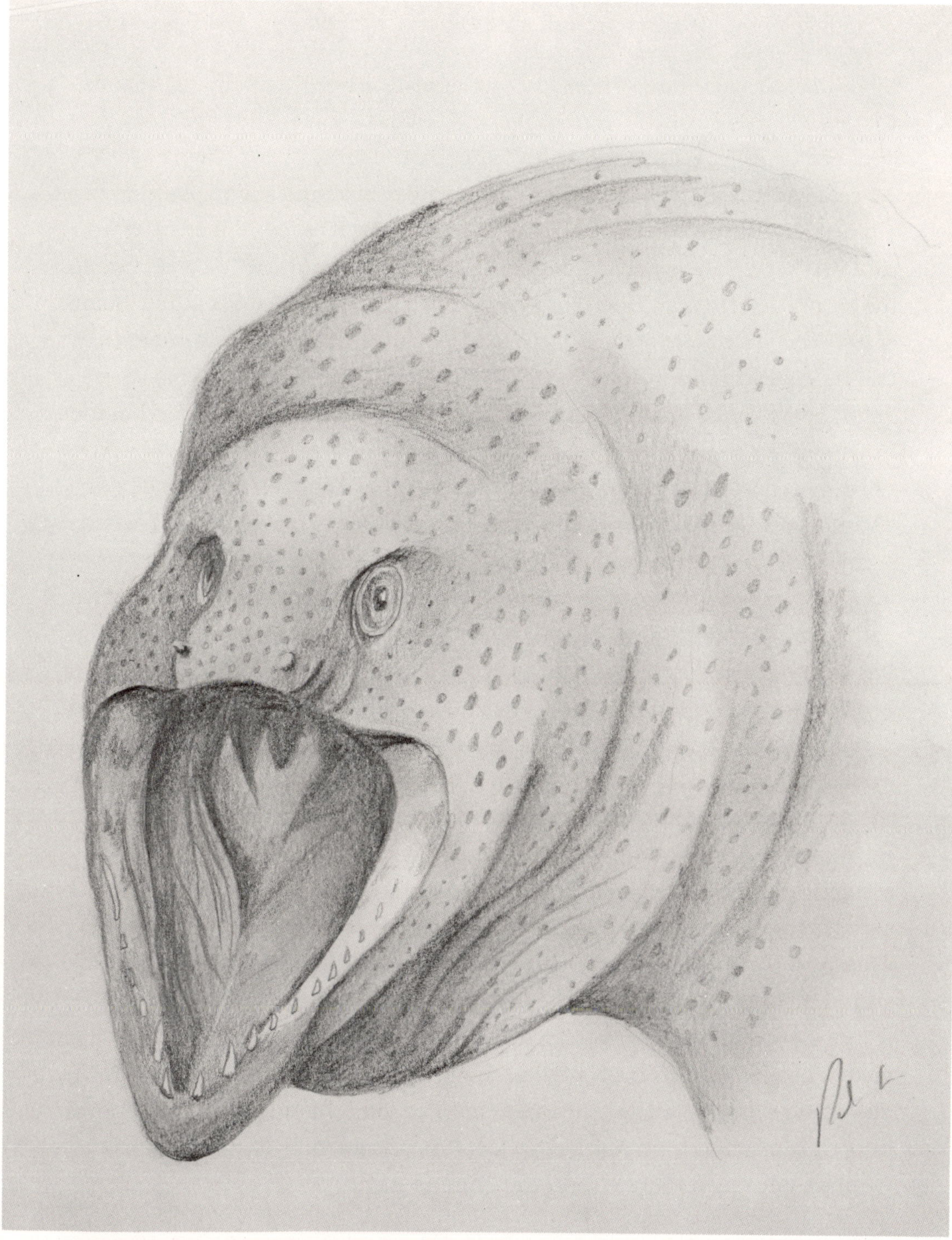

Giant Moray Eel*. This moray from Little Giftun island looks threatening but is actually expecting to be fed by divers. This artificial feeding is now discouraged.*

Eve had been a mermaid, the moray eel would have been her tempter. Not even a Giant Moray's closest friend would call him handsome. The unsmiling face and beady little eyes are hardly avuncular. Morays, and Giant Morays especially, have that distant look perfected by people who stamp passports at airports. They have an undeserved reputation for ferocity, but treated with caution they can be safely approached. The now discouraged ploy of offering the big eels food to draw them out into the open has somewhat backfired. Some of the eels now expect food and can look menacing when snaking out of their reef hideout in expectation of a tidbit. Contrary to popular belief moray eels are not poisonous. Their bite, though, can easily get infected. Morays should be treated with the respect their appearance would seem to dictate.

Respect does not involve putting the moray eel in a bottle of water and taking it home. I once found a young Snowflake Moray off Na'ma Bay in Sinai. Snowflake Morays are smaller than Giant Morays, and whereas the latter is dark brown mottled somberly with darker brown, the Snowflake Moray is white freckled and blotched with umber and yellow—rather attractive. I was watching it in the reef shallows when I became aware of a man standing behind me with a bottle of water. We exchanged niceties and then he announced his intention of taking the eel back to his hotel. I expressed concern, but he assured me it would be fine because he was going to put it in his bottle. I expressed more concern. Even without direct consultation with the eel I felt it would have severe objections, and ones that I could probably express better than it. The man seemed rather surprised. I stayed there for nearly two hours before the eel-rustler got bored and walked away. The area is now within a National Park. While taking a moray eel to a hotel room in a bottle of fresh water has always been incredibly stupid, taking anything from the water in the Park is now strictly illegal.

Back in Cairo, the city smolders through the summer. During the day it lies molten beneath a shroud of polluted smog, belching exhaust and laying down dust. The gray of the air and the concrete form a bland continuum. No creature in its right mind goes out during the day. Humans abound. Much better to wait for dusk and the relative cool of the Cairene night.

Anyone who has strolled the streets of Cairo, for whatever reason, during the hours of darkness is probably familiar with a cylindrical little animal that scurries across the road to disappear beneath a parked car or down some ill-lit alleyway. When I first arrived in Cairo, in my zoological naiveté, I pondered long and seriously over whether these shadow-like creatures were weasels or mongooses. My first caretaker, or *bawab*, a man who had trouble distinguishing dogs from donkeys, called them *nims*, the Arabic word for mongoose. Now, however, having had some experience of both beasts (and they have little in common other than general form), it is clear the elusive little cylinder is a Weasel. Experts are unsure as to exactly how the Weasel came to be in Egypt. Some believe it occurs here naturally. Others claim it was introduced, probably by the Romans. Unlike in Europe, here it is almost entirely associated

Weasel. Of all Egypt's mammal species, this is the most likely to be seen. Here it is shown in typical Cairo habitat, beneath a parked car about to dash across the street.

with humans, being particularly common in Alexandria and Cairo. Whatever its origins, it seems to be doing very nicely at the moment and has the rare distinction of winning public sympathy for its role in controlling rodent pests.

The Weasel is one of the world's most diminutive carnivores, males reaching a length of some 38 centimeters, females being somewhat smaller. In the world of the weasel, though, size is of little consequence. Members of the weasel family, collectively known as the mustelids, are unique among the carnivores in regularly hunting and killing prey far larger and heavier than themselves. When cornered they make formidable opponents. A friend of mine in Mohandiseen was once awakened from his hard-earned slumber by the sound of furious barking coming from his garden. Despite the obscene hour, fearing the worst he investigated to find his two dogs, a dachshund and a rather substantial German shepherd, with a Weasel. The irate mustelid was on its back, its teeth bared and all four sets of claws fully operational. On human intervention it beat a hurried though dignified retreat leaving a somewhat bewildered German shepherd nursing a nasty cut on his snout. Although largely nocturnal, Weasels where undisturbed will come out during the day. I once watched a male at very close quarters in one of the cemeteries in Coptic Cairo.

The Egyptian Mongoose is unlikely to be seen in the city; it is more an animal of the agricultural areas and desert margins. Despite its name, the species is not confined to Egypt but is found throughout Africa and southern Europe. While the Weasel is small, chestnut-brown above and pale below, with a slender tail, the Egyptian Mongoose is a more substantial creature. Vaguely weasel-like in form, it can reach a meter long including the heavily furred tail. The fur is coarse and grizzled gray, often with a black tip to the tail. While the weasel is valued as a predator of rodents, the Mongoose, or Pharaoh's Cat, was held sacred in ancient Egypt for its predilection for crocodile eggs and for including venomous snakes in its rather catholic diet.

It is in this latter capacity, immortalized in Kipling's Rikki-Tikki-Tavi, that the Mongoose is best known. Local 'guides' (I use the word in its very loosest sense) at the Abu Sir pyramids, whose information on the monuments is at best vestigial, claim that Mongooses regularly hunt and kill cobras at the adjoining Sun Temple. These reports have to be backed up. Suffice to say that Mongooses do eat snakes and the Sun Temple is a suitable habitat for both predator and prey.

And so to snakes. In a rather peculiar display of parental affection, my mother gave me a python to hold on my first trip to a zoo. The reptile seemed twice as long as I was, but we parted on the best of terms and I have had no fear of snakes since. In fact, I'm rather fond of them, but my affection for the animals is not widely shared. They are very useful, and though unpopular, they play an important role in the control of vermin populations.

Egypt has 34 species of snakes. Of these, the cobras are perhaps the most infamous. Myth and legend, fueled by Shakespeare (who had no background in zoological distribution and taxonomy), popularly lay the blame for Cleopatra's death on the Asp. Terribly unfair. The Asp is a small viper that has never been nearer to Egypt than southern Europe. Unless the queen imported one specially for the occasion (and the relevant permits would have taken a far longer time to process than the play itself takes to perform), the true culprit clutched to her bosom was probably the Egyptian Cobra. Egyptian Cobras can reach a length of nearly two meters and are generally a uniform brown in color, the shade varying considerably. Some individuals are marked with darker brown around the head. The famous hood is not apparent unless the animal is annoyed, whereupon it flattens the neck region and adopts its classic pose. With experience (and I can think of few people who would actually want to gain the experience), it is possible to predict the likely strike range of a cobra by how high it rears. I am not that experienced and, though an ardent admirer of the beasts, I find the use of phrases such as 'likely strike range' rather worrying in reference to a poisonous snake. Two other species of cobra occur in the country. One is the uncommon Spitting Cobra, characterized by its black neck and unusual method of delivering venom. The other is the rare Innes Cobra, only likely to be encountered in the northeast, and there only if you are very lucky.

Spitting Cobra. *The raised head of the cobra may indicate it is about to strike or, in this species, spit. However, the hood can also be raised when the animal is scared or merely curious.*

Cobras were revered in ancient Egypt. While on a purely domestic level people were quite happy to have mongooses around to keep them out of the houses, they featured widely in myth and symbol. The pharaoh wore a cobra affixed to his crown, symbolic of his rule over Lower Egypt and to afford protection. The latter role arose from one of the ancient creation myths. The enormously talented god Ra, having created himself spontaneously from the primeval waters that abounded, then created two children Shu and Tefnut, god and goddess of air and mist respectively. In the gloom that prevailed at the time, he lost his offspring and hit upon the idea of an eye to go and look for them. He thus created an eye and sent it off on its mission. During the eye's absence he created himself another eye. When eye one came back it was extremely annoyed that it had been replaced by eye two. Ra, with two eyes on his hands as it were, made eye two the eye of the moon and eye one the more important eye of the sun. He then transformed the eye of the sun into a rearing cobra in order to protect himself. Representations of Ra show him as a falcon-headed God bearing a sun disk surmounted by the cobra, while the pharaoh wore the cobra in his crown to protect himself from his enemies. Cobras were even mummified. At Tuna al-Jebel, near Minya, mummified cobras have been found amongst the millions, literally, of mummified ibises and baboons.

Most snakes are harmless, and they do not come a lot more harmless than the African Egg-eating Snake. It is a widespread species with a range covering most of Africa and parts of Arabia. In Egypt it is restricted to the southern part of the Nile Valley. Its diet is hardly catholic. As the name suggests, it feeds on eggs, and everything about it is geared to the consumption of eggs. The Egg-eating Snake is a very slender reptile, generally under a meter long, and

Egg-eating Snake. *Here shown about to engulf a pigeon egg, the Egg-eating Snake displays elaborate adaptations to its very specific diet.*

colored gray with black diamonds. It is far more slender (about the thickness of a little finger) than the eggs it eats, and has undergone some amazing adaptations to cope with its diet. The Egg-eating Snake is virtually toothless. Holding the egg against a coil of its body, it detaches its lower jaw from its upper and wraps its mouth around the egg. The muscular skin around the jaws forces the egg further down the snake to the back of the throat. Here it is prevented moving back or forward by projections on the inside of the backbone. Spiked extensions to two of the vertebrae pierce the eggshell and the egg, still held in place, is then crushed between the two holes. The contents of the egg are then free to flow into the stomach, while the shell is crushed, compressed, and ejected as a neat package.

The problem is that egg-eating has left the snake toothless and thus defenseless, so it has resorted to mimicry. The scales along the mouth are patterned so as to look like teeth. The diamond patterning of the snake resembles that of certain poisonous vipers, and when disturbed it rasps and hisses, aping the potentially lethal Saw-scaled Viper, a resident of Egypt's deserts. It even goes in for dummy strikes, lunging at the aggressor with a mouth armed with little more than gums. At least it looks good.

For those still not entirely happy about life with snakes, regardless of their usefulness and mythical significance, there is another ally. The Short-toed Eagle is a bird of prey that specializes in catching and eating snakes. While it is a common migrant, passing through in considerable numbers in spring and autumn, it is also a rather rare breeding resident. Despite its name, the Short-toed Eagle is not a true eagle. True eagles have what are quaintly known as trousers, that is to say their legs are feathered right down to their feet. Short-toed Eagles have only partially feathered legs, perhaps short trousers, and belong to a group known as the snake-eagles, named after their main prey. The 'short-toed' bit refers to their rather stunted, blunt talons. Otherwise eagle-like in form, Short-toed Eagles are about 68 centimeters long, diagnostically pale in color and with a curious, round, almost owlish head and large eyes. Thick feathering around the head protects them from their prey. Any large raptor seen hovering is likely to be a Short-toed Eagle. The only other birds of prey that hunt in this way are the Kestrel and the much smaller Black-shouldered Kite, neither exceeding 34 centimeters. With a diet of snakes, the Short-toed Eagle fills the role of an airborne mongoose. One should have sympathy with snakes. With eagles and mongooses out for a meal they have enough to worry about without suicidal queens damaging their reputation.

Cairo simmers on, it and its occupants melting in the heat. Dust seems to pervade everything, a thin beige veil over every surface inside and out. Nothing seems to shine and sparkle. Things wild and natural hardly seem to break the mold. Those creatures that stick around for the summer seem far from kaleidoscopic. The skies, even those over Tahrir, are filled with wheeling gangs of mouse-brown Pallid Swifts. Hooded Crows in funereal black and gray abound, haunting the Gezira as though auditioning for bit parts in Hitchcock's *The Birds*. The ubiquitous House Sparrow chips and chirps its way through the day, the female in neutral beige, the male hardly resplendent in black and brown. Beige is to birds what willow-pattern is to china. The mammals hardly help. Neither the Black nor the Brown rat promise flamboyance on the color front.

However, not all is as bland as it first appears. One of the plainest city birds is the Common Bulbul. It hardly grabs the eye, adding a new dimension to the

Common Bulbul. *The rather short, angular crest of the Common Bulbul is the nearest it gets to ornamentation.*

word dowdy. It's a medium sized, slim, thrush-like bird of the city gardens and is brown all over. Beneath it is paler and its head is noticeably darker, but that is as far as it goes. However, it has one of the most beautiful songs of any of Egypt's birds, a glorious, fluting, melodic bubbling usually delivered from the cover of a tree or shrub. Common Bulbuls are found throughout the Delta and the Nile Valley. In Sinai a slightly different bulbul is found. The Yellow-vented Bulbul was, until recently, considered a race of the Common Bulbul but has now been elevated to a full species. It is very similar, but its feathered sobriety is relieved by a striking yellow 'bottom' and a distinct white ring round the eye.

It would be a shame to leave the bulbuls without explaining their curious name. It purportedly comes from an old Arabic term for a small bird, the word itself deriving from the bird's call. Bulbuls have long been popular cage birds in the Middle East and, but for a probable mistranslation, might have achieved more fame than they have. In the famous *Rubaiyat* of Omar Khayyam the bird mentioned as the Nightingale, another brilliant singer colored a shade of dishwater, is more probably the Bulbul. Nightingales neither breed nor sing in the region, while bulbuls do both.

Birds and mammals apart, many other creatures are quite capable of achieving high levels of dinginess—though sometimes with a twist. For the Underwing moths, being a cryptic blend of grays and browns serves them admirably as camouflage. Against a concrete wall or a sooted tree trunk, really rather large moths are rendered practically invisible. Should some sharp-eyed predator see through the disguise though, the moth has another line of defense. As it opens its dingy forewings the hindwings are exposed. These are brilliant shades of yellow, crimson, or orange and the flash of color is thought to shock the assailant just enough to let the moth escape.

Certain species of grasshopper use a similar tactic. More renowned for their jumping ability than their flying skills, grasshoppers and their voracious relatives the locusts do indeed have wings. One of the larger grasshoppers of Egypt's deserts is a substantial insect, but, colored in muted and mottled desert hues, it is rendered almost invisible. It trusts its cryptic coloration to the point where one has to virtually tread on it before it takes flight. Then suddenly this flash of vermilion leaps up from beneath your feet. The eye is sufficiently distracted by the color that the grasshopper has time to resettle invisibly elsewhere. Reports of a hominoid moving stealthily across the desert just to the north of the ‘Ain Sukhna road, periodically leaping in the air and cursing, will be me trying to photograph the thing. Scientists call it *Oedipoda*. I have some rather more colorful terms.

Perhaps the oddest way to be really dull is to be brilliant scarlet all over. This is the curious route to hoped-for obscurity chosen by the squirrelfish and the soldierfish. A number of species of each live in the Red Sea. Squirrelfish and soldierfish are all bright red with various degrees of white marking, which should render them blindingly obvious in the reef. Red, however, is not as noticeably loud as it is made out to be, witness Cairo traffic lights. These fish are both nocturnal, spending the day hidden in coral caves and cavities. At night they emerge to feed on small fish and invertebrates. Water acts as a light filter but not uniformly. It filters out the red wavelengths of light first. Thus at depth, or in low light, anything red appears to be gray and shadowy. In the nocturnal reef the brilliant crimson squirrel- and soldierfishes appear muted and dull. Many other nighttime coral inhabitants from crabs to sea slugs are a similar hue.

Staying in the Red Sea, which, as summer moves on, seems an excellent proposition, another denizen of the coral reefs makes no pretense at gaudiness. A uniform dull brown, the Flashlight Fish appears to be a submarine bulbul without the saving grace of a good voice. However, this little fish has a large organ beneath each big eye packed full of bacteria that can produce a biochemical light or luminescence. Using special muscles, the fish can even activate and deactivate the bacteria, thus turning the lights on and off. Far from attracting predators, the Flashlight Fish uses its lights to attract the plankton on which it itself feeds.

So nothing is as simple as it seems. Dullness, like beauty, is only skin (or feather or scale) deep. Bulbuls babble beautifully, moths shock, and fish flash. Perhaps sparrows juggle.

Evening comes as some relief after a day of unforgiving heat. The city itself seems to relax. As dusk settles, a stroll along the Corniche can prove rewarding, but steer clear of the Banyan trees. Every evening they fill with twittering hordes of roosting House Sparrows, whose toilet habits are at once random and unerringly accurate. Over the river, and by the bridges especially, the air fills with thousands of bats, each one doing his or her best to keep the insect population in check. Up the river flap small flocks of large white birds, heading toward Giza in close V-formations. These are Cattle Egrets. Cattle Egrets should be familiar to anyone who has visited the Egyptian countryside. They are the stocky, white heron-like birds that can be seen following the farmers round the fields throughout the Nile Valley and Delta, waiting to see what the plow turns up. Each evening they fly to large communal roosts. The ones flapping languidly up the Nile will be heading for Giza Zoo where, along Sharia Murad, a colony has nested for decades. In breeding dress (and the birds continue nesting through July), the Cattle Egret is not the pure white it appears at first glance. Closer inspection will reveal buff on the crown, back, and belly. Not all white heron-like birds can be dismissed as Cattle Egrets. A far more elegant relative, the Little Egret, is a much more local breeder here, but it can easily be seen by visitors as a small colony nests at the Botanical Gardens on Kitchener Island in Aswan. Little Egrets are pure white, but in their breeding plumage (which by summer they should have donned) they are extravagantly ornamented in beautiful lace-like plumes falling over the back like a truncated bridal train. Their bill is longer and more dagger-like than the Cattle Egret's, and their legs are black, ending in contrastingly yellow feet.

Confusingly similar to the Little Egret is the Reef Heron. This too will now be breeding, but rather than being found inland it frequents the coastal areas, as its name suggests. A white egret stalking the reef edge along the Red Sea is almost bound to be this species. Curiously, a black egret stalking the reef edge along the Red Sea is also almost bound to be this species. Both black and white birds are forms, or morphs, of the same species, and intermediates also occur. An early riser at Sharm al-Shaykh is often rewarded by the sight of a

***Little Egret**. At one time hunted for its lace-like plumes, the Little Egret is now protected. This one, sketched in breeding finery, was seen at one of its few Egyptian nesting colonies in Aswan.*

Reef Heron picking its way along the beach at Na'ma Bay, the same beach that a couple of hours later will be packed with people.

An early morning stroll virtually anywhere along the Red Sea is likely to prove interesting. Disregarding, if that is at all possible, the likelihood of seeing dolphins swimming offshore, it is worth keeping an eye out for the birds. Some specialties such as the Brown Booby and the Red-billed Tropicbird have already been mentioned, but there are others. While the Reef Heron strides the tideline, the Caspian Tern patrols the air. This is a serious tern, the biggest of its kind. The size of a gull, it is unmistakable with its coal-black cap and bill

Osprey. *After a successful hunting foray, this bird has brought its prey back to a favored mangrove perch. Mangroves also provide Ospreys with suitable nesting sites.*

the size and hue of a carrot. On broader wings, held at a distinctive kink, Ospreys are also likely to be seen. These white and brown birds of prey feed solely on fish, which they catch in dramatic fashion, plummeting into the water to emerge with their prey clutched in their talons. Their feet are covered with specially spiky scales to cope with the slippery skin of the fish. At vantage points such as Ras Muhammad it is even possible to spot a sea turtle swimming languidly past, only to disappear with surprising speed at the slightest disturbance. By far the most likely to be seen is the Green Sea Turtle.

For those unlucky enough not to make it down to the Red Sea, or to the Mediterranean coast for that matter, it is sadly still possible to see sea turtles. One might trip over one on the sidewalk on 26th July Street in Zamalek plonked outside a fish shop, or spy one languishing in a pet shop. The hapless reptiles are normally displayed in a washing-up bowl of fresh water, sometimes with bits of lettuce-leaf that serve better as a wreath than as food. Sea turtles supposedly enjoy the full protection of Egyptian law.

Autumn

Egypt slips rather than plunges into autumn. The trees, for the most part, do not go in for the dramatic leaf-shedding of their more northern counterparts. The weather teases—a few cooler, breezy days may seem to promise the end of a sweltering summer, only to be punctuated by another heatwave.

In Europe, the natural world starts bracing itself for winter far earlier. Come August the first migrant birds are heading south, very sensibly, to overwinter in tropical African climes. By September, the exodus of birds, having taken full advantage of Europe's seasonal plenty, is well under way and Egypt's airspace is once more full of birds in transit. As in spring, it is the largest birds that capture the imagination and attention—the vultures, eagles, cranes, pelicans, and storks. Their large size and broad wings limit them to soaring flight over long distances and a migratory route south dictated by thermals, the rising currents of warm air that they ride on.

The storks are among the earlier migrants. The White Stork is a stately bird, a meter long, with scarlet legs and dagger-like bill. As the name suggests it is largely white, with black on the wings. At first glance it might be confused with the white egrets, but on land it is generally larger and more substantial in build, the leg and bill color immediately distinguishing it from the Great White Egret, and in flight it holds its neck out straight. The herons and egrets always fly with the neck tucked up. White Storks may be encountered in wheeling flocks

Migrating birds of prey. Here a single immature Steppe Eagle can be seen soaring amid a small flock of Steppe Buzzards. Flying so high, these birds might appear to the naked eye as mere specks in the sky.

along the Red Sea coasts, in Sinai, or in the Nile Valley. From mid-August through to mid-September they are common migrants, and when not in flight may be seen patrolling the fields. A good place to find storks is in the irrigated areas north of the Cairo–Isma'iliya road. They are impressive birds, tall, strutting the fields with an almost military bearing (two of their Indian relatives are called Adjutant Storks). Storks are firmly embedded in European folklore, popularly associated with the delivery of babies. While none have been seen bearing human infants while passing through Egypt, they do have a place in Egyptian lore too. The Bedouin and villagers from Upper Egypt believe that the rings taken from the legs of storks (placed there by ornithologists) are lucky charms.

The White Stork, being white, is aptly named. It comes as no surprise therefore that the Black Stork is largely glossy black, relieved only by a white belly. It shares with its white relative the scarlet bill and legs. Black Storks are much rarer, more solitary birds, unlikely to be seen in large flocks. Rather, they can be found in ones and twos amongst the hordes of raptors working their way south. They tend to migrate later than the White Storks, the first ones likely to be seen in September.

Impressive though the storks undoubtedly are, it is the cranes that I most look forward too as autumn approaches. From September through to the middle of November flocks of these magnificent birds ply their way south. Having gained height by ascending on currents of warm air, the cranes form distinctive V-shaped flocks, wings flapping in a deceptively leisurely fashion. From below their flight outline is similar to that of the stork, cruciform with the long neck extended in front and the legs behind. The Common Crane is a uniform gray in color though, relieved only by black, red, and white markings on the head. Watching a flock of cranes pass overhead is more than a visual experience, it is a heartmoving one. Cranes call constantly as they fly, a deep,

Common Crane*. Seen here on migration, the Common Crane is one of my favorite birds. Flocks of migrating Common Cranes may contain much rarer species.*

rather sonorous *krook, krook, krook*, a sound as deeply associated for me with spring and autumn as the horn is with Cairo traffic. The constant resonant cries seem to be somewhat organizational, keeping each member of the team in contact with the others. Every now and then the lead bird at the head of the V changes, the order shuffles, or the end of one arm of the V peels off and forms a new small V of its own, all the time the cranes calling, calling, calling. Migrating storks seem to fly in a chaotic jumble. With cranes there is a definite order and apparent organization.

There are plenty of reasons for watching migrating Common Cranes (for most of us aesthetic ones—they are beautiful), but even the serious ornithologist who may have ticked off *Grus grus* (for that is the Latin name by which serious ornithologists know the bird) on their list would do well to watch the passing flocks carefully. A smaller bird, similarly gray but with a black neck, flying with the Common Cranes may prove to be the rare Demoiselle Crane. As dainty as the name suggests, it is a mere meter long, compared to the 130 centimeters of the Common Crane. Interestingly, both species are well represented in friezes in a number of Egyptian tombs. In the tomb of Ti at Saqqara there is a scene showing cranes being force-fed, indicating that the ancients may have kept them at least semidomesticated. So accurate is the craftsmanship of these friezes that both species can be clearly recognized. There is one curious feature, however. Cranes (in common with all birds) have a traditional complement of two legs. It is the generally accepted norm from which no crane knowingly deviates. In some of the pharaonic representations of flocks of cranes this maxim is broken. Find a group of cranes in a frieze and the chances are that the number of legs falls short of the two-per-bird rule.

Just as in spring, there can be no better place to witness the migration of the large birds than by the Red Sea. In fact autumn is even better than spring. After several months of constant burning sun, the water has warmed up and lost the cold edge that made it a little chilly earlier in the year. A heart uplifted by cranes passing aloft can be further uplifted, or plain stopped, by the attractions in the coral reefs beneath the water—not least the sharks.

To most people sharks are to be avoided. They have a certain reputation, fueled by rumor and exaggeration and inflated by films such as *Jaws*. If *Jaws 1* had people fleeing from the water, then *Jaws 2* kept them firmly landbound, and just when they thought it was safe to go back in the water, the movie-makers preferred they went back in the cinema and came up with *Jaws 3*. *Jaws 3* was different from the others because Bruce, the plastic stand-in for a real shark, with teeth of rubber and a latex hide, could be seen munching away at swimmers in 3-D. Now for some home truths. Although hundreds of millions of people work or play in the sea each year, there are on average only 50 recorded attacks by sharks, of which ten may be fatal (compare this with the 700,000 sharks killed in driftnets in 1990 by the Japanese alone). Most of these attacks are by Great White Sharks. The Great White Shark has never been reliably reported from the Red Sea. No diver or snorkeler has ever been killed by a shark in the Egyptian Red Sea. The very few fatalities have been swimmers, who splash and create a disturbance, or spearfishers. Spearfishing is illegal in Egypt, and spearfishing in waters with sharks is rather like setting up a barbecue in a pride of lions. For the thousands of divers who visit the unparalleled Red Sea reefs each year, one of the biggest excitements is seeing sharks.

My first dive with sharks was disappointingly unmemorable. I came back from a trip to Sinai able to tell people that apparently I had dived with sharks. I say apparently because, unlikely though it may seem, I was totally oblivious to them, and it was only when I clambered back into the boat that I realized what I had missed. There was, according to my fellow divers, a stream of Black-tip Reef Sharks and Gray Reef Sharks passing by. Some counted seven, others eight. My count was a round zero. To add insult to injury, one of the Gray Reef Sharks had the curious distinction of being the largest fish that I have so far managed not to see. Those who did get a look estimated it to be some two meters long. I take their word for it.

The Black-tip, the Gray, and the White-tip Reef Sharks are the three species most likely to be encountered by divers because, as their name suggests, they inhabit the reef areas, the White-tip especially occurring in surprisingly shallow water. Larger species, such as the formidable Tiger Shark and the gentle, plankton-eating Whale Shark, occur in deeper, more open water.

It might seem somewhat curious that anyone, let alone a naturalist—who should have more finely honed observation skills than most—can keep the company of two-meter sharks and be completely unaware of it. The key lies in the rest of the reef. Such is the glorious, kaleidoscopic complexity of the reef community that on every growth of coral, new dramas are being constantly played out. Take as an example my favorite, the cleaning station.

Fish, despite their watery medium, get dirty. They suffer from external parasites. Old food sticks to their teeth. Fish, like humans, need to be spruced up every now and then, the scales polished, the fins wiped down, the mouth freshened up. Some sea creatures manage to clean themselves. The Pelagic Sea Snake ties itself in a knot and then runs its body through the knot, wiping off unwanted clingers-on such as barnacles. (There are, however, no sea snakes in the Red Sea. Reports of such reptiles invariably relate to morays or, more likely, to species of serpent eel.) Fish are generally of less flexible build. The fin, though admirable as a means of propulsion is neither dexterous nor flexible enough to double as a toothpick. Instead, a fish feeling in need of a clean will attend a cleaning station. These specific areas of reef are manned by Cleaner Shrimps and/or by a little fish called the Cleaner Wrasse.

The wrasses are a diverse group, closely related to the parrotfishes. Wherever you turn in the reef you are likely to see a wrasse. The rainbow-colored fish in checkered pinks and blues, so confiding that it almost knocks on the mask of a snorkeler or diver, is Klunzinger's Wrasse. The two-meter-long Napoleon Fish, in the past often fed hard-boiled eggs by divers (a practice now discouraged) is a wrasse. The Birdfish, with its long beaklike snout, is a wrasse too. It is a family of bewildering variety, its members often of appreciable size and flamboyantly ostentatious coloring. The Cleaner Wrasse, though, is neither large nor gaudy. It is a diminutive little fish, scarcely ten centimeters long, slender, and striped lengthwise in black and pale blue. It

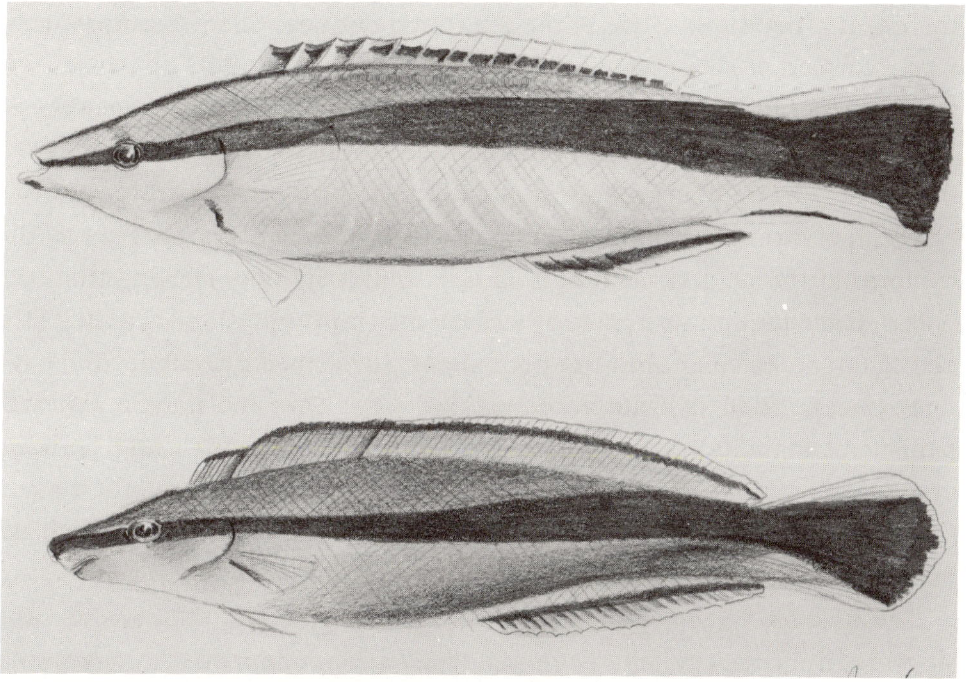

Cleaner Wrasse and Mimic Blenny. *The plate shows the clear similarity between these two unrelated species of fish, a similarity from which only the blenny, below, benefits.*

cleans much bigger fish. Its clients, the moray eels for example, are often potential predators, for whom a Cleaner Wrasse would make a tasty snack. Yet moray eels need their teeth cleaned as much as any other fish. The solution reached is a behavioral one. A fish approaching a cleaning station and wishing to be cleaned will behave in a specific manner. The wrasse's response is to perform a set 'dance' that inhibits any aggressive behavior on the part of the client. The wrasse can then go about its business in total safety. The fish being cleaned will let the wrasse enter the mouth to clean between the teeth and will even allow access (lucky old wrasse) to the gill cavities. It seems a fairly straightforward relationship. The big fish gets cleaned, the little wrasse gets plenty of food scraps. Then enter the Mimic Blenny. The Mimic Blenny is a fish of equally diminutive proportions as the Cleaner Wrasse, and is almost identical in shape and color. It too performs the wrasse's dance, thus pacifying the client. The Mimic Blenny, though, is no cleaner. Client becomes dupe when, instead of cleaning, the blenny nips off bits of fin or flesh with razor sharp teeth. Mimic Blennies are far less common than the Cleaner Wrasse. Were they to be anywhere near as numerous as the fish they impersonate, then the client fish would cease to respond to the behavior of the wrasse.

Cleaner Shrimps perform a similar role to the Cleaner Wrasse. They hardly conform to the popular idea of a shrimp. Infinitely more striking than the boiled, pallid pink thing popularly awash with thousand island dressing on a bed of lettuce, Cleaner Shrimps are brilliant red, striped with white, and have long, slender, and delicate antennae and legs. They too have a series of behavioral adaptations to ensure that the fish wanting to be cleaned does not end up with an easy meal. Cleaning stations are fascinating areas of the reef to watch. Apparently they are so fascinating that other more immediate attractions—namely two-meter sharks—can be missed.

The Cleaner Wrasse and its clients enjoy a symbiotic relationship, one where both partners benefit. Such arrangements are common in the coral reef. While the Cleaner Wrasse has a steady stream of different clients, the clownfish forms a permanent bond with its rather unlikely partner, the sea anemone. The Twobar Clownfish is an attractive fish some ten centimeters in length and goldfish-like in build, but rounder from the side. It is orange-brown in color, with two white bands on the side of the body, one just behind the eye, the other nearer the tail. Close inspection (and the delightful thing about this little fish is that even the most timorous of snorkelers can get close) shows that the bands are bordered in electric blue. Clownfish are invariably seen in and around sea anemones—an unusual haunt, for fish the size of the clownfish constitute a perfectly acceptable meal for a sea anemone. The tentacles of the anemone are armed with poison cells used to capture and paralyze prey such as small fish, and hence small fish normally give the anemone a wide berth. Not so the clownfish. It not only swims round and through the waving tentacles with impunity but on the approach of danger actively seeks shelter within them.

Twobar Clownfish. The clownfish is seen here engulfed in the poisonous, Medusa-like tentacles of a sea anemone. It's an unusual haunt, but the association benefits both animals.

Clownfish are not immune to the stings of the anemone. The secret lies in mucus. Young clownfish, through rubbing themselves against the anemone, gradually cover themselves in the mucus that covers the tentacles. Once the fish has assumed this slimy mantle, the anemone is fooled into accepting it as part of itself, and so does not even try to sting it. Scientists with a particularly mean streak have tried wiping the mucus off the clownfish. Sure enough, the hapless fish gets stung. The clownfish clearly benefits from its relationship with the anemone. It gains protection from larger fish and at night is probably one of the sounder sleepers in the reef, cocooned safely in its protective Gorgon's head. The duped anemone might miss out on one meal but seems to benefit by feeding on the leftovers of the food the clownfish captures. This unusual couple are a common feature of the reef. Rightly confident of their anemone minder, clownfish can be approached closely and watched (a fascinating pastime, though snorkelers should remember to come up for air every now and then). Get too close and the little clownfish may attack. They get very defensive about their anemone home.

Not all creatures get on quite so well. The caretaker's cat turned up with an interesting, if rather reluctant playmate one day. Striding out of the heat, I was about to spring up the steps to my flat when I noticed the white and marmalade feline, which normally lies curled up and comatose on a small chair that must now be his by right of occupation, in a state of unprecedented activity. He was leaping, dodging, springing, and cuffing, and behaving very like a cat is meant to behave. This, from an animal that spends his normal day doing a very passable imitation of a soporific cushion, demanded investigation. He turned out to be playing with a small lizard, a lizard that I rescued before the game became dinner. It was an Ocellated Skink.

For those unfamiliar with the Ocellated Skink, it is a lizard with the build of a slightly flattened cylinder, with a conical snout and a long whiplike tail that gives it a length of some 14 centimeters. I doubt if this was this particular skink's first encounter with a cat. Its tail was paler than the rest of it and

Ocellated Skink. *This particular individual was rescued from a one-sided play session with a cat and released, none the worse, into a nearby garden.*

disproportionately small. Like many lizards, a skink can shed its tail when caught. The disconnected appendage will continue to twitch away while the rest of the skink escapes to grow a new but smaller tail. Skinks are distinctive for their very smooth, shiny skin, an adaptation to their way of life spent largely underground. In the Ocellated Skink the skin is a pale olive-brown and covered in white centered black spots, or ocelli, that give the creature its name. This species and the Bean Skink are two of Egypt's commonest lizards.

I feel the Bean Skink's name requires some kind of explanation. The Bean Skink does not look like a bean. Neither does the Bean Skink eat beans. It can pass its natural span without ever coming across a bean. However, a Victorian naturalist translated one of its local Arabic names as 'the lizard of the green beans'—possibly because it was common in the fields—and the name stuck. The breeding male is a rather splendid creature with upperparts dark brown, often with five longitudinal cream stripes and a striking yellow-orange patch running from nostril to shoulder. The female is duller but more distinctively striped, and the immature has a vivid blue base to the tail. Bean Skinks seem particularly common in agricultural areas and can often been seen on the mud banks of irrigation ditches.

A third species of skink, Audouin's Sand Skink, exhibits another trait typical of skinks. They have a tendency to loose their legs. They have not done this in the same way that they shed their tails of course, but many are so adapted to life underground, often in sandy areas, that they have evolved into forms with much reduced or absent limbs. Audouin's Sand Skink has only vestigial forelimbs and not much larger hind ones, giving it an almost snakelike form. This is ideal for movement in soft sand where a swimming motion is more effective than running. Indeed, a fourth species of skink is called the Sandfish.

As for the Ocellated Skink, I released it unharmed in a nearby garden, where it should be stalking its insect prey. The caretaker's cat is short of a playmate, but it's a resourceful creature. After a few days' sleep something else will probably come along.

Reptiles must have mixed feelings about autumn. On the one hand they all, from the humblest gecko to the mightiest monitor lizard, get complete legal protection from October through to April. This is to ensure that they are not hunted during their hibernation period, when they are torpid and hence more vulnerable. Popular they might not be, but valuable they are as important pest controllers in town and country. Some of the rarer species, the boas, the sea turtles, and the crocodile, for example, are protected year round. However, while reptiles may gain legal protection, hundreds of thousands of potential natural predators are passing through the country between August and November. Many birds of prey will take reptiles on occasion and some, such as the Short-toed Eagle, are specialists. Storks, cranes, and herons will all snap up a reptile if the opportunity arises.

One of the commoner migrants, and one of the most opportunistic, certainly not above plucking a skink from a field of green beans should it get the chance, is the Black Kite. The Black Kite is a medium-sized bird of prey, fairly uniform dull brown, with rather slender wings and a long, slightly forked tail. It lacks the size and power of the eagles, the speed of the falcons, or the elegance of the harriers, but for all this it shows supreme mastery of the air.

On migration, the slender wings and the shallow fork of the tail distinguish the Black Kite from any other bird of prey. However, it is also a resident in Egypt. It is a common bird in Cairo and can be seen over most parts of the city, even the very center, gliding above the smog on slightly bent wings. The river is a good place to spot them, as the banks seem to prove profitable hunting grounds. Outside the city they are less common. During recent years they suffered greatly from the widespread use of pesticides, and their population fell. With more control on agro-chemicals, they seem to be making a comeback. Black Kites are basically scavengers, feeding on whatever they can find. They can be extremely bold. Some years ago in Kashmir (Black Kites have a wide range), I lost a sandwich to a Black Kite that swooped down and snatched it off my table while my back was momentarily turned.

Black Kite. *Together with the Kestrel, the Black Kite is Cairo's most common bird of prey. The pale bill shows this bird to be an Egyptian resident.*

Migrating Black Kites, merely passing through, can be readily told from the resident birds by their bill color. Egyptian Black Kites are of the African race and have a yellow bill. Migrating kites have black bills. This small detail is quite noticeable at close quarters, especially if the bill is wrapped firmly round your sandwich.

Another common bird that is both resident and migratory is the Swallow. The Swallow is the familiar, aerial bird with the glossy blue upperparts and deeply forked tail. It is common throughout much of Egypt, including Cairo,

Pallid Swift. *As autumn advances some Pallid Swifts migrate south. Others, such as these seen over old Cairo, remain here throughout the winter.*

where it can be told from the Pallid Swift by its more deeply cleft tail, slower flight, and shorter wings. Swallows seem completely at home with humans. They nest on our buildings, hawk insects over our towns and villages, and even drink from our swimming pools. Swallows do not land to drink but glide just above the water surface and scoop up water with their bill. An arrow of indigo sweeping across the pool is likely to be a thirsty swallow.

As with the Black Kites, resident swallows differ from the European migrants sensibly fleeing oncoming winter and heading for the African tropics. Local birds have rich rufous underparts. The visitors have a rufous throat and a dark blue breast band but otherwise creamy underparts. Ornithologists are currently trying to make a case for the Egyptian Swallow to be elevated to a separate species. As things stand at the moment they are merely a race.

While Swallows spend most of their time hawking insects in the air, I have seen them catching insects on the ground at the desert margins. The ground is not their preferred habitat though. Their legs are small and weak and totally unsuited for walking. When they do land it is generally to perch on telegraph wires, a nice gesture to the birdwatcher. Telegraph wires are splendid open, easily watched perches that render their underparts clearly visible.

Courtship is not on the minds of many animals at this time of year, but the Hyrax is an exception. Found in the deserts of Eastern Egypt and Sinai, the Hyrax is a curious beast. The size of a large rabbit, it looks like a rodent, but its closest relatives are the elephants and the sea cows. It lives in colonies in the mountains, feeding on whatever vegetation it can find. Around September the dominant males in their rocky territories become increasingly aggressive, defending a harem of up to twenty females. For the hyrax, mating is a brief, vigorous affair, and for such a small animal the pregnancy is long—nearly eight months. Early writers warned against eating the hyrax on the grounds that it 'cheweth the cud but cloveth not the hoof.' More to the point is that the Hyrax is normally packed full of a variety of highly unpalatable internal parasites.

It would take a trip to the remoter parts of the desert mountains to find Hyraxes, but autumn is a good season for urban mammals as well. With many

***Hyrax**. The Hyrax is one of Nature's more improbable creations. The size of a large rabbit, there is little from its outward appearance to suggest that its closest relatives include the elephants.*

trees and plants in seed it is little short of a bonanza for the rodents. One well worth watching out for—at least I think so—is the Black Rat. The Black Rat is generally not actually black, more of an ashy brown. It differs from the larger and often darker Brown Rat in its larger eyes and ears and longer, more slender tail. I think it is a rather attractive animal, but I fear others may still be put off by the fact that it was a carrier of the Black Death and indirectly responsible for the deaths of millions of people in days gone by. The tables have now turned. Poisons, better sanitation, and control have meant that the Black Rat has disappeared from many of its old haunts. Though it is now thoroughly at home in Cairo, it is not a native. It was probably introduced by trade

Black Rat. *These rats are common but generally shy. This one, though, was munching away fearlessly on the seeds of the Kurrajong tree while I watched it from a balcony barely three meters away.*

caravans from India, or through the ports. Indeed, it is also known as the Ship Rat.

In the field of natural history, naming a creature can be a fickle and misleading business. A fine example of the vagaries of nomenclature is afforded by the Egyptian Plover. This is a strikingly marked bird, cream and gray, patterned boldly with glossy black and white. Anyone unfamiliar with the Egyptian Plover might from its name leap recklessly to the conclusion that the bird was, firstly, a plover and, secondly, found in Egypt. Wrong on both counts. The Egyptian Plover is more closely related to a group of aberrant wading birds known as the pratincoles and coursers than to the true plovers. And, sadly, it is no longer found in Egypt, the last definite sighting north of the Sudanese border being back in 1937. The reasons for its disappearance are probably related to the changing regime of the Nile. Nesting on sandbanks exposed after the inundation, the Egyptian Plover found fewer available sites as the waters came under increasing human control. This, coupled with increasing human disturbance, saw their numbers dwindle and finally disappear. A rather more fanciful explanation comes from its alternative name, the Crocodile Bird. This arose from its reported habit, described first by no less a person than Aristotle, of entering the mouths of basking crocodiles to pick away at the food debris accumulated between their teeth. As the crocodile disappeared from the Egyptian Nile below Aswan (the last was one reported in 1891) so did its avian dentist. Since the crocodiles seem to be staging something of a recovery behind the High Dam, perhaps the Egyptian Plover might too return. Until then, one ornithologist recommended calling it the Non-Egyptian Non-Plover, rather cumbersome but more accurate.

It would be a delightful bird to have back in Egypt. The Egyptian Plover provided me with one of my biggest thrills in Khartoum, Sudan. In the middle of the Blue Nile, just before it meets the White Nile at Omdurman, lies Tuti Island. Off the eastern tip of Tuti lie extensive sandbanks that seemed perfect habitat for the Egyptian Plover. I spent a whole day watching the banks. Other birds, flocks of ibises and kites and Spur-winged Plovers, came and went and

***Egyptian Plover**. The name of this incredibly dapper little bird is most misleading. This pair was observed off Tuti island in Khartoum.*

then, just when I was about to leave, my patience was rewarded. A pair of Egyptian Plovers appeared at the water's edge. When one relies on field guide illustrations for one's picture of a bird, the real thing can sometimes fall short of an over-glamorous portrait. Not so the Egyptian Plover. I have never seen a smarter, more clean-cut bird. Not a feather out of place. The pale underparts were crossed by an immaculate black breast band. The white eye stripe positively glistened against the glossed black head, and the gray upperparts looked velvet-smooth. When they lifted their wings in display or annoyance the black and white patterning was stunning in its clarity and boldness. It was a bird worth finding. My last day in Sudan was rewarded with a flock of nine, the same number as the cloud I was on.

Unfortunately, the Egyptian Plover is not the only species lost to Egypt in recent times. The Sacred Ibis, presently not found north of Sudan, is rather like an egret in build, with a white body, long black legs, a long, unfeathered black neck, and a long, black downcurved bill. It carries a short train of black lacy feathers on its rump, much as a Victorian lady carried a bustle. Often depicted in tomb paintings and reliefs, the Sacred Ibis was revered by the ancients as the living image of Thoth, the god of wisdom. Immortality was of little consequence to more recent ibises, and the last one was recorded in Egypt back in 1891 (not a good year for wildlife, with the crocodile disappearing as well). At one time they must have abounded. At Tuna al-Gebel, near Minya, there are catacombs where three and a half million mummified ibises have so far been discovered. This has left archaeologists with something of a dilemma. Killing a Sacred Ibis was a sufficiently heinous offense in ancient Egypt to carry the death penalty. Yet here were literally millions of neatly prepared, trussed up and embalmed ibises. It is unlikely that any ancient in his right mind would have had the time or inclination to follow a chosen ibis around until it dropped dead of natural causes. One theory points to a priestly racket. The Thoth clergy, ever anxious for more wealth, farmed the ibises, and when laypeople felt the urge to venerate the god they could buy one ready prepared from the priests, avoid the death penalty, and hopefully leave their chosen deity suitably chuffed.

Sacred Ibis. *However much this bird was revered by the ancient Egyptians, it is doubtful whether even they would have described it as beautiful.*

Thoth not only took the form of an ibis but was also depicted as a baboon, the Hamadryas, which is still also known as the Sacred Baboon. Not content with accumulating prodigious numbers of dead ibises, the custodians of Tuna al-Gebel also filled the catacombs with the mummified remains of some half a million baboons. Alas, the Sacred Baboon, too, is no longer found in Egypt, but it still occurs in Ethiopia, Somalia, and the Arabian peninsula.

It seems unfortunate and rather ironic that so many of those animals worshiped by the ancients are extinct in modern Egypt—deification seems to have done them little good. But while the ibis, baboon, crocodile, and hippo (the goddess Taweret, associated with birth) have disappeared along with the belief in the gods they represented, others have survived. The mummified remains of both the domestic cat and the Swamp Cat have been found in their hundreds of thousands at Tell Basta near modern Zagazig—both are still around in modern Egypt. The God Horus was often represented by the falcon, several species of which live and breed in the country to this day. Finally, there is Anubis, 'Lord of the Mummy Wrapping,' commonly represented as a jackal or a man with the head of a jackal.

The jackal is an opportunistic, carnivorous canine, a close relative of the domestic dog, from which it is often difficult to distinguish in the field. In general form it is not unlike a German shepherd, but slighter, with a coat of grizzled black and yellow hair and a distinct shaggy mane. The tail is relatively short, thickly haired, and often tipped black (compare with the Red Fox, much smaller with a white-tipped tail). Despite persecution, Jackals seem to have escaped the fate of most former gods and remain relatively widespread, though secretive and nocturnal.

However, the Jackal's popular identification with Anubis has been questioned. The City of Anubis, its site near present-day Asyut, was known under the Ptolemies as Lycopolis, literally 'Wolf City.' Ibrahim Helmy, who spent over twenty years researching and co-authoring his work on Egyptian mammals, maintains that there are not, nor have their been, any wolves in Egypt, a logical conclusion since they are at best stragglers from Arabia to neighboring

Palestine. Further, he claims that Anubis was in fact a fox. Looking at the famous black statue of Anubis in Hall 45 of the Egyptian Museum in Cairo, one can be convinced by his arguments. The God's features are certainly vulpine. The face is thinner and more pointed than a Jackal's, the ears relatively larger and the tail far too long and bushy. The Anubis question is a difficult nut to crack. I reserve judgment. I have yet to see Jackals in Egypt, though I have seen them in sub-Saharan Africa, where they are called Golden Jackals to distinguish them from other jackal species. I doubt whether many people are particularly concerned. Anyone about to be embalmed can scarcely be in a fit state to worry whether the process is being supervised by a wolf, a jackal, or a fox.

As autumn advances, the birds seen passing over Egypt change. Each species seems to have its own timetable. Some, like the Black Kites and the Steppe Buzzards, can be seen throughout the season. Others, such as the gaudy turquoise Roller and the Woodchat Shrike, may have been common in spring but are much less so in autumn. Yet others pass through sporadically. Adult Cuckoos leave Europe well before their young, who will have been reared by foster parents such as Willow Warblers or Meadow Pipits. The newly fledged young migrate south later, and with no experience and no guidance from their parents find their way unerringly to their African wintering grounds. It is one of the many mysteries of migration.

For many birds food, or rather the lack of it, will provide the trigger to migrate. Where the food supply runs down gradually as winter nears in Europe, the birds will head south gradually. Species that rely on one prey item alone will often head south en masse once their prey species disappears. Such a bird is the Honey Buzzard. Another example of a misnomer, the Honey Buzzard feeds on bees and bee larvae rather than honey, and is not a buzzard but is more closely related to the kites.

Honey Buzzards are well adapted to their prey. The legs are heavily scaled and the head and neck very densely feathered against the stings of bees and wasps. In flight they are longer-winged and longer-tailed than buzzards, with a smaller, almost pigeon-like head and a barred tail. It might take only a couple of cold spells in northern Europe to decimate their prey populations. Unable or unwilling to switch diet, they will then all head south at pretty much the same time. One week the skies can be devoid of Honey Buzzards. The next there may be droves, although rather smaller droves than in spring.

Perhaps a small note on names may be valuable here with regard to birds of prey. Americans do not speak the same English as the English do, and nowhere is this more apparent than when the two nationalities go birdwatching. In the States a Buzzard is a large, vulture-like bird that the English call a Turkey Vulture or Black Vulture. However, the Black Vulture in America is a different and unrelated bird to the European bird of the same name that is a true vulture. What the English call a buzzard—the Steppe Buzzard and Long-legged Buzzard in Egypt—the Americans would call a hawk. Thus in the States an English birdwatcher would call their Red-tailed Hawk a Red-tailed Buzzard. The English use the word 'hawk' for birds such as the Sparrowhawk (a visitor to Egypt) and the Goshawk (a rare winter visitor to Egypt). Curiously enough, so do the Americans, but then they also apply the name Sparrowhawk to the American Kestrel, a close relative of the Common Kestrel found in Egypt and more properly a falcon. Easy. It is this type of confusion, even between those who supposedly speak the same language that drives scientists to use Latin names.

It is unfortunate that as the birds of prey fly over Egypt many seem to find their way into the animal markets and petshops of Cairo and elsewhere. Kestrels, Black-shouldered Kites, Honey Buzzards, various eagle species, Lanner Falcons, and Marsh Harriers have all been seen for sale at one time or another. Neither is this trade restricted to birds. Fennec Foxes, sea turtles, Long-eared Hedgehogs, and Ibex have all found their way into these sordid markets. Despite much of this trade being illegal, the traders openly flout the law. On one memorable occasion a gazelle was being offered for sale along Qasr al-'Aini Street. The creature was imprisoned in a cage barely larger than itself, placed out in the street within inches of the traffic. On top of the cage housing the terrified gazelle was perched a captured Lesser-spotted Eagle, tied to the cage by a piece of rope. Passers-by, curiosity aroused, were poking and prodding both animals. It took several weeks to get the animals released—if a life spent in Cairo Zoo can be rightly called released.

Occasionally the trade backfires and the animal victims come out on top. One or two species have managed to build up self-sustaining wild populations from escapees. Parrots are birds normally associated with the tropics, with dangling vines, lush jungles, and the odd pirate. Cairo, though deficient in at least the first two, is the adopted home of one species, the Rose-ringed Parakeet. This is a popular cagebird throughout its extensive African and Asian range, though I do not know why. I had the dubious honor of looking after one for a couple of weeks, at the end of which I was close to avicide. Oh, they are pretty birds, slim, long-tailed, and soft green in color with a contrastingly crimson bill. The male has a black throat patch and a pink collar, hence the name. Visually pleasing, it is a bird for the incurably, totally, stone deaf. Its voice is a shrill piercing shriek, delivered at full volume when it is guaranteed to annoy the most. This particular creature would sit all pretty, soft and green and quiet, until the phone rang. Then up he would start. Friends calling thought they were phone witnesses of a particularly gruesome murder. Two weeks with this monster and I am not in the least surprised that these birds have managed to establish themselves in the wild here. My only quibble is with calling them escapees. I am sure the several hundred pairs that now breed in and around Cairo did not escape. Their cage doors were probably swung right open and their chastened owners were probably down on bended knee begging them, pleading with them to leave, to fly off anywhere, preferably a long way away. No one knows exactly when the parakeets took up residence in Cairo, but it was probably in the late nineteenth century. Today, ironically, the best place to see them—yes, and hear them—is Giza Zoo, which has a healthy feral population. Watch the sky for a streak of green borne with rapid beats on pointed wings. Then wait for the shriek.

Another cagebird that has successfully established itself in the wild is the Avadavat. A native of India, the Avadavat is not a true urbanite but has

Golden Oriole *(male). This bird is most likely to be seen in large gardens and fruit orchards. The male, pictured here appropriately perched in a mango tree, is one of Egypt's most brilliantly colored birds.*

established colonies at the barrages north of Cairo, at Gebel Asfar just outside the city, and in many areas of the Delta. Less spectacularly tropical than the parakeet, it is nevertheless an attractive bird, despite being a close relative of the lowly House Sparrow. The size and build of a small finch, the breeding male is crimson red, dotted with white. The female and juvenile, as seems to be their lot in the bird world, are duller. Unfortunately, at this time of year the male looses some of his breeding splendor, molting and donning his more subdued winter plumage.

If, when you are traveling around the fields in autumn, or indeed throughout the winter, some of the lowly House Sparrows seem to look a little less lowly than normal, then they may well be Spanish Sparrows. The Spanish Sparrow is a common migrant, and many remain here over the winter in the Delta and along the Nile Valley. Male Spanish Sparrows have white cheeks, a chestnut crown, and far more black on the breast and flanks than the male House Sparrow. The females, too, have more heavily streaked flanks than their House Sparrow counterparts. Despite its name, the Spanish Sparrow is rare in Spain. When a Spanish Sparrow crossbreeds with a House Sparrow—it seems ornithologists at times are better at telling the two apart than they themselves are—the result is a hybrid known as an Italian Sparrow. This is a misleading name, since it is unlikely to be terribly common in Italy—the Spanish Sparrow, a required parent, is not found there.

Another bird worth keeping an eye out for during an autumn stroll, particularly in orchard areas and gardens, is the Golden Oriole. Some birds are somberly colored but command attention because of their brash, forward behavior or loud, musical or otherwise voices. The male Golden Oriole is stunningly colored and yet very difficult to see. He is a thrush-like bird, some 24 centimeters long, and a brilliant citron yellow in color with contrasting black on wings and tail. The female is duller and greener. Despite its plumage, and being not uncommon, the Golden Oriole is difficult to find amongst the dense foliage of the fruit trees it favors. Not surprisingly for a bird so beautiful, it is frequently represented in ancient Egyptian tomb paintings. However, many of the representations show the birds being trapped and netted. They appear to have been considerable pests, eating large quantities of fruit. They are reportedly still trapped today, though they are unlikely to be serious crop thieves.

The control of supposed pest species is a tricky business, not least in identifying the culprit. Fruit bats eat fruit. However, this fruit (and they have a particular penchant for bananas) is grown not for fruit bats but for human consumption. A neighboring Middle Eastern country decided that since fruit bats ate fruit that should be being eaten by humans, they would eliminate the fruit bats. They did this by fumigating the bat roosts with toxic chemicals. While this did make a minor dent in the fruit bat population, it decimated far larger numbers of insect-eating bats that shared the roosts. With the insect-eating bat numbers far lower, the insect pest populations mushroomed, necessitating the widespread and previously unnecessary use of insecticides. Furthermore, it was then shown that the fruit bats had no impact on the commercial banana crop. Fruit bats have no interest in unripe fruit. They like their bananas overripe, sticky and sweet. Since bananas for human consumption are picked while still green, the bats were merely eating those fruits left on the trees after the harvest.

The same bat, the Egyptian Fruit Bat, is found, not surprisingly, in Egypt. It is by far the country's largest bat, with a wing span of over half a meter. With the tail totally surrounded by the flight membrane, it looks something like a flying tea towel. Size alone should identify it at night. Unlike the much smaller insect-eating bats, fruit bats have large eyes. At night, passing a torch over a tree of ripe fruit may reveal feeding fruit bats as pairs of glowing orange eyes reflected in the light. Fruit bats are found throughout the Delta and the Nile Valley, including many parts of the city.

My flat seems to look out over one of Cairo's chief 'batways.' On warm evenings they flit past in bewildering numbers, many perhaps heading for the river, where the insect pickings are richest. Bats' habits are hardly conducive to easy identification. Their twilight emergence, their rapid, flitting flight, and the fact that there are more species of bat in Egypt than of any other type of mammal all conspire to make life very tricky for the crepuscular naturalist. In near darkness it may not be immediately apparent whether one has seen a Mehely's Horseshoe Bat or a Lesser Horseshoe Bat in rapid flight, when the only significant difference is in the shape of part of the nose. Unless the enthusiast is the unlikely owner of a bat-detector, a box of electronics for lowering the frequency of a bat's high pitched sonar, the flitting little beast, long since disappeared, will merely be put down as a generic bat. Not all is lost.

***Egyptian Fruit Bat**. Looking at the vulpine features of the fruit bat, it is easy to see how it got its alternative name of flying fox.*

In the city, probably the most common bat, Kuhl's Pipistrelle, can be distinguished by its small size when flying with other species. Conversely, the much rarer Free-tailed Bat can be told by its large size and substantial tail. Other differences are more subtle and need experience. The Naked-bellied Tomb Bat, common around the Pyramids, can be told in the hand by its hairless underparts. In the air, its fast, direct flight is subtly distinctive. The Egyptian Slit-faced Bat, on the other hand, has a slower, more erratic flight.

It is during the autumn that the insect-eating bats prepare for winter. Some, such as the Lesser Rat-tailed Bat, build up stores of abdominal fat over the late summer and autumn that allow them to remain active through the winter, when their insect prey is much less abundant. Large numbers of this species can be seen around Dendera temple near Qena and in the past also at Qasr Qarun in the Fayoum. On my last visit to the latter, the large number of desiccated dead bats suggested the monument had been sprayed. No doubt the insect pests in the area are delighted. Other bats, such as Kuhl's Pipistrelle, do not lay down fat stores, but become less active during winter. Pipistrelles do not hibernate in the true sense but will continue to appear on warmer evenings throughout the winter. The flypasts outside my balcony are more humble than those reserved for heads of state (jets being considerably noisier and faster than bats), but I enjoy my batwatching evenings.

Bats are not the only Cairene creatures active and aloft at night. The Senegal Thick-knee's cry can be heard piercing the night air throughout the city with something approaching hysteria. Also at large is the Barn Owl, whose ghost-like form can be seen silently passing through the night like a giant white moth. The rounded wings of the Barn Owl bear it in total silence. As with most owls, the wing feathers are softer and more loosely barbed than most birds' and edged with hairlike filaments. These effectively deaden sound as the owl flaps its wings on the downbeat, eliminating the whoosh or whistle that one hears with ducks and geese for instance. Silent flight has two advantages. Firstly, the prey (generally rodents in the Barn Owl's case) cannot hear the owl approaching. Secondly, owls rely more on sound than sight when hunting.

Feathers. *Top to bottom: Hoopoe (left: secondary; right: scapular), Barn Owl (secondary), Long-eared Owl (primary), juvenile Egyptian Vulture (tertiary).*

Most birds, like humans, have earholes that are symmetrical. One either side, and in more or less the same place. Owls do not. Their ears are askew, one high on the head, one low. In human terms it would mean one ear level with the mouth, the other with the eyebrow. This arrangement, hidden beneath the feathers, is thought to help in locating sound direction in pitch darkness. If the owl were to make a noise while flying it would not only scare off its prey but would also disrupt its own hearing.

The Barn Owl is a pale bird, gold above, subtly speckled with silver gray, and snowy white below. At night it appears all white. Most distinctive is its heart-shaped facial mask punctuated by dark eyes. In winter, a less common visitor is the Long-eared Owl, a darker, heavily streaked and vermiculated owl, with long ear tufts. Both are useful birds. Their pellets reveal that they seem to subsist mostly on House Mice, supplemented by the odd sparrow. Unfortunately, owls too find their way into the pet markets and shops. The Barn Owl in the illustration was seen for sale at the biggest pet market, at Sayyida 'Aisha near the Citadel. The bird was clinging to the side of its spherical cage in the absence of even a perch. The rusting metal hoop through the wing was to stop it from flapping around. The price was 15 Egyptian pounds. Barn Owls are protected under Egyptian law. Cage bars are not the type of 'protection' envisaged.

While city mammals such as the diminutive pipistrelles may be getting somnolent, in remote corners of the Eastern Desert and Sinai the rutting season of the Nubian Ibex continues through to November. Brought to the brink of extinction, the Nubian Ibex, a species of wild goat, still exists in small numbers in these mountain regions. It is an imposing mammal. The male can reach 80 centimeters at the shoulder and weigh 90 kilograms. He is grizzled pale brown, with a white belly and black markings on the flanks, forelegs, muzzle, and back. The female is smaller and less strongly marked. The main cause of their demise has been overhunting. They are, apparently—as I hasten to add I have not taken part in this 'sport'—rather easy to hunt. Unlike many desert mammals they need water regularly and are creatures of habit. All the

***Barn Owl with doves.** A portrait of abject misery from one of the pet markets.*

Nubian Ibex. *During the rut, the male ibex, seen here in the foreground, uses his massive backcurved horns to battle rival males in competition for the much smaller females. The rocky terrain is typical ibex habitat.*

brave and intrepid hunter has to do is to ensure he is at the water hole at the right time of day, generally early morning or dusk, wait for them to arrive, and then blast away. The results can be seen all over the walls of the Manial Palace, in the shops of Kerdasa, and in the souk at Aswan. Only a few wadis still reverberate with the sound of clashing ibex. The males, with huge, backswept horns, hurl themselves at each other in an effort to win domination over harems of the much smaller females. The kids will be born in spring.

Fortunately, the Nubian Ibex is now a protected species, and recent reports indicate that it may be present in the National Park at Ras Muhammad. There, at least, it should be safe, as should the Green Sea Turtle. These reptiles are thought to breed on certain beaches within the park, now out of bounds to visitors. The female sea turtle comes up the beach at night and digs a hole with her paddle-like forelimbs. She then deposits her eggs within the hole and covers them once more with sand. Undisturbed nest beaches are crucial to the survival of all species of sea turtle. The Green is the most frequently seen of those species occurring in Egyptian waters. It has a smooth, highly domed carapace, or shell, and can sometimes be seen paddling languidly on the surface of the sea, periodically craning its neck, like some reptilian periscope. At any sign of danger it will disappear with a speed not normally associated with near relatives of the tortoise. A rarer species, the Hawksbill, can be distinguished from the Green by the serrated margins of the carapace and its less domed appearance. It may breed on some of the remoter Red Sea islands.

I have seen Green Sea Turtles at Ras Muhammad, not only from the water but from boats and, on one occasion, from the sea cliffs. Indeed, even fishwatching can be done from the comfort of a boat or the beach. There seems to be a general consensus that fish belong in water. Certainly, water is the preferred medium for most fishes, but there are some nonconformists, not least the Flying Fish. There is some controversy over whether Flying Fish actually fly or merely glide. Whatever, a first encounter with them is a weird experience. What looks like a shaft of silver suddenly appears out of the waves and flies like an arrow some considerable distance before re-entering with barely a splash. It all happens very quickly. Then another, and another. It takes a bit of getting used to when it is explained that these shafts of silver are in fact fish.

Other fish also leave the water, though they do not 'fly' but leap out of it and crash back in a display known as breaching. Halfbeaks, sizable fish with the bottom jaw greatly elongated and the top not, are prone to this, emerging from the water like a great silver cutlass. Manta Rays, too, breach. The same shape as stingrays, the body flattened to form broad wings, they can reach seven meters across. Seeing a Manta breach must be like watching an aircraft emerge from the sea.

Most fish that leap out of the water, however dramatically, fall straight back in. One, though, is more than happy to leap out and stay out. Along the rocky shores of the Red Sea lives the Leaping Blenny. It is a small, slender fish, some seven centimeters long, cryptically colored in beige banded with dark brown. The eyes are large and bulbous, perched on top of the head. The Leaping Blenny would probably go largely unnoticed were it not for the fact that it spends most of its time out of the water. It is most often seen resting on rocks right at the water's edge. On being disturbed it leaps into the sea, but it can survive for long periods out of water. It has enlarged gill chambers that can hold water while it is on its chosen rock and can even breathe through its skin. It is often very common, and a crowded rock can almost look as though it is effervescing if the fish are disturbed.

One of the special habitats preserved at Ras Muhammad is an area of mangrove swamp, one of the northernmost in the world. The mangrove is a tree specially adapted to life in brackish water, even water as saline as that of the Red Sea. It can therefore support itself in areas most other plants would find totally inhospitable. The mangrove swamp at Ras Muhammad supports a small population of the Green-backed Heron. This bird has a range that extends over much of Africa, but it is very local in Egypt. It can easily be distinguished from any other heron by its small size and dull green-gray coloration, relieved by a dark green-black crown. The Green-backed Heron is

***Green-backed Heron** (foreground) and **Goliath Heron**. These two species, size-wise at opposite ends of the heron spectrum, are both inhabitants of the Red Sea mangrove swamps.*

a shy, secretive bird, spending most of its time skulking around the dense mangrove thickets. With a little patience, however, one might see it nervously emerging from its mangrove cover to search for small fish and invertebrates in the surrounding shallow lagoons.

A rare breeder in Egypt's southern Red Sea mangroves that disperses north in autumn is a heron built on a very different scale. The Goliath Heron is a meter and a half long with a bill more like a pile-driver than a dagger. Size alone should identify this, one of the world's largest herons, but its chestnut crown and neck and purple-brown upperparts further distinguish it. It is a rare bird in Egypt and merely a straggler to the northern Red Sea coasts. Anyone seeing it, though, is sure to want to know what on earth it is. It is huge.

Winter

Winter hardly hits Egypt with Arctic ferocity. However, for Cairo's human residents the onset of winter comes as a welcome relief after the furnace-like heat of the Egyptian summer. For a hominoid, the drop in temperature means a few months respite from the grinding cycle of cold drinks and copious perspiration. Our response is to reach to the wardrobe for a sweater and to turn down the fan a notch. The mind boggles as we realize what a flexible, adaptive creature we are.

The drop in temperature, particularly at night, and the shorter days force certain other animals to change their habits. While there is no need for any beasts to adopt the more extreme winter survival ploys of their northern relatives, such as turning white or donning a heavy winter coat, many show more subtle adaptations. The easiest is simply to move. Blessed with the power of flight, birds can up and off with the first cold spell. By December the summer visitors, such as the flamboyant Blue-cheeked and European bee-eaters, are well ensconced in their tropical African wintering grounds. And very sensible too.

Others will have followed suit. Most Egyptian Vultures, for instance, become non-Egyptian vultures as they soar off to sub-Saharan wintering grounds. Hardly one's usual idea of a vulture, and certainly not as grand or imposing as my young hotel-mate the Griffon, these dirty white, slender-billed

Egyptian Vulture*. Graceful aloft, the Egyptian Vulture seems to belie its alternative name of Pharaoh's Chicken.*

raptors are also known as Pharaoh's Chickens. It's unflattering but not a complete misnomer. On the ground, they do bear some resemblance to a bald, baggy-trousered White Leghorn. Aloft, they are transformed. Like other vultures they are broad-winged, and though small by vulture standards, with a span of a mere two and a bit meters, they demonstrate complete mastery of flight. In the air they are unmistakable because of the striking black and white wing pattern and the distinct diamond-shaped white tail.

Egyptian Vultures are unusual in being one of only a very few birds known to use a tool. Having arrived in their East African wintering grounds, the vultures may well hanker after an Ostrich egg, a nutritious and substantial meal, but one unfortunately enclosed in a formidably thick shell. For the brighter vulture this presents no problem. It takes up a suitably sized stone and uses it as a hammer to crack open the shell. Across the other side of the world, the thorn-wielding Galapagos Woodpecker Finch and, in Papua New Guinea, the paint-daubing Satin Bowerbird are among the very few other avian tool users.

Such behavior has never been demonstrated by Egyptian Vultures in Egypt. This may be because nobody has really bothered looking, but is more likely to be because the Ostrich is so rare in Egypt now. I have seen photographs of them from the deserts southeast of Aswan, and they probably occur in Gebel 'Elba, but these areas are so hard to get to that there has been little direct observation. Ostriches probably breed within Egypt's southern borders some years, a number of females laying eggs in a single large scrape that is then guarded by the male. They were not always this rare. Rock carvings from places such as Silwa Bahari in Upper Egypt show Ostriches, as well as a number of mammal species such as elephant and rhino. These carvings, dated to around 6,000 years ago, indicate that Egypt was rather cooler and wetter than at present, supporting a typical grassland/savanna fauna. Throughout the pharaonic period, Ostriches were represented in tomb paintings, sometimes in scenes of Ostrich hunts, sometimes (as at al-Bersha near Minya) in scenes that indicate a form of domestication. The birds were common in Egypt well into more modern times and were reportedly still hunted extensively in the late eighteenth century. Hunting pressure ultimately proved too much, though, and the ostrich disappeared as a regular breeding species at the end of the last century. Since the 1960s, however, there have been regular reports from the south of the country and evidence of breeding. The Ostrich could well be back, but only in small numbers. Any Egyptian Vulture wanting to flaunt its precious tool-using skills would still be far better off farther south.

As Ostriches stand over two meters tall and weigh up to 155 kilos, it is perhaps just as well that they do not fly. All other Egyptian birds do, though.

Greater Flamingo. *Seen here in the saline shallows of Lake Qarun, Greater Flamingos are among Egypt's most exotic winter visitors. The small, black birds are Coots.*

Many follow the Egyptian Vulture south. By November the screeching flocks of Pallid Swifts over the city center will have thinned as the less hardy of their number will have left. Anyone anxious for a last glimpse of the Rufous Bushchat will have missed their chance come winter—no amount of stomping through their favored summer farmland haunts will produce a Bushchat after November. They have all fled south. So have many birds that have reared young here, including the Turtle Doves and Sand Martins.

Replacing them, though, are thousands of other birds, different species, which have bred in Europe and for which Egypt serves as an important wintering area. Along the Nile Valley, and on the Delta lakes, gather large flocks of waterfowl—Mallard, Teal, Pintail, Garganey, and Shoveler. In Aswan, a stroll behind the Old Dam on the track to the Philae ferry can often be rewarded by mixed aggregations of Pochard and Widgeon. Ducks are not the only visitors. A trip to the brackish lakes at Wadi Natrun can yield small parties of Black-necked Grebes and larger numbers of waders such as Ruff and Redshank. However, it is Lake Qarun in the Fayoum that can boast the most impressive selection of winter guests. The lake has over recent winters become home for a flock of up to 370 Greater Flamingos.

Lake Qarun is a large, shallow lake, the shape of a rather elongated oblong. It used to be freshwater, but the changing regime of the Nile has meant that the water has become increasingly saline, with associated changes in the species of fish found there. The increasing brackishness has created problems for the local farmers—a recent flood rendering several plots of land unusable—but has had little effect on the birdlife. What did affect the birds was excessive hunting. Since hunting was banned on the lake in 1990 the bird population seems to be recovering. It is impossible not to see something special on Qarun in winter.

The big attraction is the flamingos. Grotesque in the most gracefully attractive way, the flamingos on Qarun are not the bright crimson birds of most zoos. How pink a flamingo is depends on what it eats. People expect pink flamingos, so zoos feed them such things as grated carrot or brine shrimp to turn them pink and keep the visitors happy. In real life, some flamingo populations are really that pink, notably those in the West Indies. But most are not. Locating flamingos in the Fayoum, one should look for flocks of white birds that appear pale pink only at close range. Once found, they are unmistakable, with incredibly long, pink legs, an incredibly long and sinuous thin neck, and that large, bent beak that Alice found so useful as a croquet mallet. The flamingo feeds by scooping up beakfuls of lake-bottom detritus and then squeezing out the water with its exceptionally large tongue. The highly nutritious soup that remains is a mixture of algae, diatoms, and small mollusks and crustaceans. We would call it sludge, but the birds thrive on it.

Seeing the flamingos at Fayoum is one of the most dramatic wildlife sights Egypt can offer. Watching a flock pick their way elegantly through the water on stilt-like legs is as exciting as it is unexpected. Against the austerely dramatic backdrop of the arid cliffs of the Western Desert, these birds, looking so beautifully tropical, become almost surreal. Too close an approach and they flag. The heads rise from the water, their slender necks erect and straight, and their bills all point in the same direction like a series of pennants on slender poles. Then they may take flight, pattering as one across the water before finding the air. Now they become pink. The deep salmon of their wings and the contrasting black primaries are exposed to add to the drama as the flock wheels off. It's unforgettable. For once, Nature can truly be viewed through rose-tinted binoculars.

The attractions of Qarun are not confined to the lake itself. The farmland to the south of the lake is a good place to find two of Egypt's more common birds of prey. The Black-shouldered Kite is arguably Egypt's most attractive raptor. It is small, 34 centimeters long, snowy-white below and soft gray above, with black on the 'shoulders' (in birds what appears to be the shoulder is in fact the wrist). Even at some distance the large, piercing red eyes are apparent, frozen in an icy glare. Seemingly more common than it was a decade ago, the Black-shouldered Kite is easy to see, as it normally perches openly on exposed tree limbs, the tops of bushes, or, most frequently, along telephone wires. From

Black-shouldered Kite. *This is one of Egypt's most common birds of prey, its most elegant, and one of the easiest to observe.*

***Black-shouldered Kite.** This raptor can often be seen hovering, scanning the ground for small prey items.*

these open perches the birds go on hunting sorties, hovering when they spot something on the ground that may represent a possible meal. The only other hoverer of this size is the Kestrel. The kite is much paler and shorter-tailed, with blunter wings, white from below with the black showing through as sooty gray. Competent though they are, they do not seem to master hovering to quite the effortless extent of the Kestrel.

The other hunter of the Fayoum is harder to find but also frequents telephone wires. The Little Owl is a small owl with a disproportionately large head that itself has disproportionately large, yellow eyes. The head is rather square and flat-topped, with white brows. The upperparts are brown, spotted white, and the underside white, streaked brown. All owls are hard to find, but the Little Owl is slightly easier than most. Like the kite, it hunts from perches, though rather than using telephone wires it uses the poles themselves. Where the wires meet the horizontal brackets sticking out from the pole there is a spindle through which the wire passes: two wires, two spindles. Two wires, three spindles, and

***Little Owl**. This diminutive owl is shown in a typical haunt, perched against an old wooden telegraph pole. As in all owls the eyes face the front, giving it a poor field of vision. To compensate for this owls can, as here, twist their necks through more than 180 degrees.*

the chances are that the third spindle is a Little Owl sitting absolutely motionless on the bracket of the pole. On discovery it will turn and glare with eyes that make a Black-shouldered Kite's look warm and welcoming. If alarmed it will bob up and down and then fly off with a distinctive undulating, bounding flight.

Back in the center of Cairo, changes are apparent. There are fewer bats around, as they switch their metabolic processes down a notch, some going into a state of torpor during the cooler spells. Reptiles, too, respond to the temperature drop. While the bats tuck themselves away in darkened nooks and crannies, turn the metabolism down, and wish winter, at least in part, good-bye, the geckoes join me inside. I have no objection. The most common urban species, the Fan-footed Gecko and the Turkish Gecko, both thrive on insects and are infinitely more interesting to have around than a can of aerosol. Furthermore, though I have never been intimate enough with a gecko to check, they seem to smell a lot less unpleasant than a spray.

While I welcome the geckoes in my apartment with open arms, my cat has reservations. On the one hand, they are entertaining. On the other, they make her look stupid. Geckoes share with skinks the ability to lose their tail in moments of crisis. The tail remains twitching for a period after detachment, while the tailless reptile scuttles off to grow another. On several occasions I have found my cat staring intently at a faintly twitching tail while a gecko, generally a young one of little experience, is halfway up the wall staring intently at the cat. It's a little disappointing that Nature's most perfectly developed predator, her most superbly adapted killing machine, can fall for such a ruse.

Unlike geckoes, cats cannot run up walls, which is probably where geckoes should safely stay. Contrary to popular belief, geckoes do not have suckers on the ends of their feet. Their ability to cling to vertical surfaces, even glass, arises from the structure of their toes. Most have flared toes, the underside covered in large, plate-like scales. These scales are covered by thousands of microscopic hairs that are themselves branched and that can grip the most minute of crevices. To a gecko a wall, or even a ceiling, is anything but smooth.

***Fan-footed Gecko**. The unusual structure of their feet enables this and other gecko species to cling to the smoothest surfaces. This door would be relatively rough terrain for the gecko.*

Less noticeable—and luckily so, for also less welcome—visitors from the outside are mice. There are two species of mouse resident in Cairo. The House Mouse is a true cosmopolitan to be found virtually throughout the world, though largely courtesy of humans. Its real home is probably Turkestan. It is thought to have first reached Egypt centuries ago via the old trade routes and caravans. The other species, the Cairo Spiny Mouse, is a true native. Despite its name it is not restricted to the city but is found throughout Egypt (though not in true desert), most of Africa, and the Middle East. Of the two, the Cairo Spiny Mouse is the more likely to be found in the city, cheerfully sharing the houses and offices of human residents. These urban mice are generally much darker than their country cousins, a common phenomenon among all sorts of urban animals. The Spiny Mouse is some ten centimeters long, with an equal length of tail, prominent

***Cairo Spiny Mouse.** This is a variable species. The individual shown here, with its uniform, dark coloring, is a city dweller.*

ears, and a series of flat spines along the back. The lighter House Mouse is smaller, with proportionately smaller ears and, should you get close enough, a paler belly. As winter closes in it is more likely to seek the warmth and food of human dwellings but is generally unable to compete with its larger relative.

Geckoes most people can handle—though probably not literally—and most of us are unaware that we support a rodent population. However, a cold spell can produce more extraordinary house-guests. A couple of winters ago I was woken one morning by the phone. It was a friend who calmly announced that he had a snake in his kitchen and could I come and have a look at it. I was delighted because, at variance with the rest of the human race, I like snakes. I told him I would be around as soon as possible. He told me there was no hurry, he would 'clean round it'. I got round there some twenty minutes later. The reptile in question was coiled tightly round the bottom frame of a Butagas cylinder. This probably explained how it had happened to be in an eighth-floor Zamalek apartment, having come up unnoticed beneath the Butagas cylinder. The narrow head, long, slender body, and distinctive cream and brown longitudinal stripes identified it as an African Beauty Snake, and as such harmless. I untwined it, and after a brief photo-session found it temporary accommodation in a pillowcase. We then drove out to the countryside beyond Imbaba and released it in some farmland where, hopefully, it is still helping control the rural rodent population.

Such encounters are unusual. Snakes generally shun publicity. However, coincidentally, my nearest approach to a car accident was due to the same species of snake. A few kilometers north of Meidum, heading toward the pyramid, I was forced to swerve, narrowly missing an irrigation ditch, by an African Beauty Snake unexpectedly crossing the road. Had I ended up in the canal I might well have been the first person to have caught bilharzia, albeit indirectly, from a non-poisonous snake.

While unlikely to choose eighth-floor apartments in Zamalek, many of Egypt's reptiles seek out warm and sheltered crevices in which to spend the winter. All reptiles of any species enjoy complete protection under Egyptian law in the winter months between October and April. This is in deference to their value as pest controllers.

Not all is movement. While dusk on the balcony proves rather less bat-rich in winter than in summer, one sound transcends the seasons notably—the rising hysterical shrieks of the Senegal Thick-knee. The pair that nested near me in the summer are still around, filling the evening air with their manic cries. Some evenings it seems as though the streets are alive with Thick-knees. In the bird books, which face the onerous task of trying to reproduce birdsong in words, the call is generally transcribed as *kvi, kvi, kvi, kvi, kvi, kvi* . . . and so on. This is not overly helpful, but the cry of the Thick-knee should be familiar. Taking *kvi, kvi* . . . etc. as a starting point, imagine an urgent, piercing

Senegal Thick-knee. *This peculiar wader is now a common urban bird in Cairo, but its distinctive voice is far more often heard than the bird is seen.*

crescendo of staccato cries that start slowly, rise in pitch and tempo, and then fade again. If you happen to be in a foul mood, it sounds just like mocking laughter. The perpetrator of this row is the Senegal Thick-knee. Delivering its cry under cover of dusk or darkness, the bird invariably goes unseen despite being a common city resident.

Thick-knees are a small group of wading birds, also called Stone Curlews or Dikkops, whose rather quirkish name comes from their swollen leg joints. The Senegal Thick-knee is typical of the group, a sturdy, streaked brown bird with longish, yellow legs, a short straight bill, and huge, staring, yellow eyes. The large eyes are indicative of its nocturnal habits. It is an African bird, and the Egyptian population represents its northernmost outpost. Generally found near water, the Senegal Thick-knee naturally nests on sandbanks. However, in Egypt, and in Cairo especially, it has become an urban bird finding the flat city rooftops a convenient sandbank substitute. Now thoroughly at home, even in central Cairo, its piercing cries have become a familiar, though hardly tuneful, city sound. Egypt's other thick-knee, the Common Stone Curlew, is very similar in appearance but different in voice. It is a bird of dry, open country and farmland.

Another African bird for which the Nile Delta is the northernmost limit of its range is the Senegal Coucal. Though active by day, the coucal is also a bird where the sound is more familiar than the sight. Its 'song' is a series of deep, sonorous, hollow notes often rendered as *who, who, who, who, who, who* and continuing for up to a score of *whos*, deepening and quickening. The Senegal Coucal is a shy bird of thick cover, but once seen it is absolutely unmistakable. About the size of a small crow, it has a long, graduated, black tail. The upperparts are rich chestnut, the crown and nape black, and the underparts cream.

More striking than pure color, though, is its general demeanor—what ornithologists call its 'jizz'. The jizz of a bird is the indefinable collection of features that make it that bird rather than any other—its stance, its mannerisms, its general feel. The Senegal Coucal always looks as if it has thrown its feathers on in a hurry and has not had the time for even the most

***Senegal Coucal**. A bird of thick cover, the coucal, seen here in a mango tree where it has just landed with characteristic clumsiness, is an unusual relative of the Cuckoo.*

cursory of preens. At no point is this more apparent than when it has just glided from one piece of thick cover to another. It lands as though it has never done it before, crashing into the foliage, its chestnut wings at sixes and sevens and its tail loosely splayed out at some unnatural and quite ludicrous angle. It perpetually looks as though it has had a shockingly late night. Although coucals are relatives of the cuckoos, they do not share their relatives' parasitic habits and they rear their own young. Though shy, the Senegal Coucal is a fairly common (though strictly rural) bird throughout the Delta, but rarely seen further south in Egypt.

While some animals are shy, bashful, or plain elusive, others are rather forward. I was recently approached by a young Common Mole Cricket. I am not

Common Mole Cricket*. Mole crickets are aptly named. Here the huge, shovel-like forelegs can be clearly seen. The male can also sing.*

averse to being approached by young mole crickets, indeed I find them impressive and interesting insects, but to attempt an introduction on the tennis court was foolhardy on the insect's part. Luckily, on this occasion, I spotted it early and managed to skip nimbly around it during a hard-fought rally. She (even young mole crickets can be sexed easily and quickly by those eager to do so) was returned to a more natural habitat soon afterward.

The Common Mole Cricket is aptly named. It is related to the grasshoppers and crickets and shares the subterranean habits of the mole. It is a bulky insect, some six centimeters long, and has powerful, flattened, shovel-like forelimbs, with which it burrows through the soil. Like the crickets, the mole crickets go through a series of molts, called instars, before reaching maturity at some two years of age. At each successive molt the wings grow larger, so that age is fairly easy to determine. The adult is a robust beast, mole-gray to brown in color, with the forelimbs and portions of the other two pairs of legs densely haired in a rufous-gold velvet. While mole crickets do some damage to root crops through their burrowing, this is more than offset by the number of insect pests they eat. Male mole crickets sing, or rather stridulate. In order to stridulate, they rub their forewings together. A series of studs rasps across the wing veins, producing a sound described as a 'low pulsating chur.' While the Common Mole Cricket is relatively numerous in the Delta, its relative, the African or Dwarf Mole Cricket is a rare and more diminutive insect.

While on the subject of insects, I received a phone call once from a worried friend. He had found a Praying Mantis on his washing line and was at his wit's end as to what to do with it. It was not eating the lettuce that it had kindly been offered. From the description it transpired that the mantis was a young one, since its wings were poorly developed. Young or old, no self-respecting mantis will touch lettuce. I naturally asked whether it had come with any of its brothers and sisters, and suggested that it might be more interested in these. The Praying Mantis will feed on any insects it can capture in its fearsomely spiked forelegs, but is especially into cannibalism. The female lays several score eggs, and the first hatched tend to feed on their kin as they emerge into the big, wide, and for them rather brief world. Unfortunately, this mantis was a loner, perhaps having already consumed its peers.

Cannibalism in mantises is not limited to the young. Courting can prove a dangerous, if not suicidal, exercise for an amorous male. Compared with the female he is rather puny and very sensibly makes his advances tentatively. If the advances are not tentative enough, then he gets eaten. If he is successful in wooing his formidable intended, the two will then mate. After mating he is then eaten. Its a no-win situation for the poor male, one that has lead to the mantis being proposed as the symbol for women's liberation. The closest relative of the mantis is the cockroach, various species of which are found throughout Egypt. In Cairo the Common, American, and German cockroaches all abound.

A recent jaunt up the Nile revealed a creature normally very reluctant to show itself. In describing the Purple Gallinule the bird books all employ adjectives such as 'elusive', 'retiring,' and 'secretive.' The gallinules south of Beni Suef have clearly not read the guides and are unaware of how they should be behaving. At times they seemed to be coming out of the reedbeds and parading around in the open in a manner that verged on the ostentatious. Possibly as winter progresses cover gets thinner and food gets harder to find, so forcing the birds into the open. Possibly the birds have just realized quite how stunning they are to look at and decided the time has come to flaunt themselves. Whatever, once in the open, Purple Gallinules command attention. While they share the general form of the Moorhen, or Common Gallinule, they are far larger, and deep purple-blue in color. The form breeding in Egypt has a green back. The legs, feet, and huge bill and frontal shield are all bright crimson. The tail is held cocked up, revealing the white rear. It was with this part of the bird that I was most familiar. A glimpse of gallinule generally means a flash of white undertail disappearing into frustratingly dense marsh vegetation, where the purple renders the bird almost invisible. Yet here, along the Nile banks, the gallinules were positively flirtatious. They appeared in their twos and threes, and on one memorable occasion in a group of ten. Purple Gallinules are shy where persecuted, and this willingness to appear in the open is a positive sign, especially given the density of human population. Their future looks bright as well. With the completion of the Aswan High Dam, their reedbed habitat has increased along the Nile Valley and the birds have re-appeared in many places they had vacated, not least the Fayoum. The only two birds that a Purple Gallinule could be confused with are the Moorhen and the Coot. Both are much smaller. The Moorhen has a red bill with a yellow tip, yellow-green legs and a white streak down each flank. The Coot, a common winter visitor, is uniform black, relieved only by a white bill and frontal shield. It is less restricted to prowling the banks, and can be seen swimming along with flocks of ducks.

Purple Gallinule. *Normally shy, at certain times of year the Purple Gallinule can be seen stalking out in the open, particularly in areas flanking dense reedbeds.*

Other birds appear up the Nile in winter. Large flocks of waterfowl congregate and can be seen bobbing around on the river or resting on exposed sandbanks. An exposed sandbank gets an ornithologist's heart all aflutter. Ducks, geese, waders, gulls, and plovers of a myriad different species can turn up on an exposed sandbank. On this particular trip, Gray Heron were present in large numbers. The Gray Heron is a typical heron, long-necked and long-legged, with the usual dagger of a bill. The Gray Heron is the largest of Egypt's common herons, with a gray back, white underparts, white neck, streaked black, and a long black crest. Many of the individuals to be seen along the river in winter are less clearly marked and generally duller. These are first winter juveniles. The Great White Egret is only slightly smaller, at just under a meter. It can also be seen along the Nile in winter, where its large size and pure white plumage make its identification easy.

Sandbanks can produce surprises. Scanning one with my binoculars just south of Luxor one winter I was rewarded with a small flock of Egyptian Geese. Egyptian Geese are really oversized ducks, largely brown and orange, with a glossy green wingbar and a dark eye patch. Restricted as a breeding species to Lake Nasser, they winter further north and are not uncommon around Luxor. The real surprise on the sandbank, though, was a single White Pelican. Flocks of pelicans are quite common on migration, huge white birds with black-bordered wings, flying with the head and ample bill tucked in. In winter they are rare. Perhaps this one had decided it was an oversized, pallid Egyptian Goose, and as such Luxor was a quite normal place to spend the winter.

Herons and their like may be large, and the gallinules impressive, but my favorite bird of the Nile is the Lesser Pied Kingfisher. It has a curious distribution in Egypt, breeding in the north of the Delta and then again from around Sohag south. In winter it can now be seen virtually along the length of the Nile and along the larger irrigation canals too. There is little 'lesser' about the Lesser Pied Kingfisher. It has a huge black dagger of a bill, and at 25 centimeters is considerably larger than the European Kingfisher, a winter visitor. It is a short-tailed, black and white bird with a spiky crest. The

Lesser Pied Kingfisher. *The striking patterning and distinctive hunting techniques of this species make it one of the most noticeable birds along much of the Nile.*

underparts are white with a single (in the female) or a double (in the male) black breast band. However, the Lesser Pied Kingfisher draws attention to itself less through appearance than through behavior. Like many of its tribe, it is a fish-eater. It catches its prey by hovering up to 30 meters above the water, the black dagger pointing downward, and then plummeting down once it spots its prey. Should the fish see the kingfisher coming and take evasive action, the kingfisher may abort in mid dive and sweep up once more to resume its hover. Once prey is caught, the kingfisher flies with it low and fast over the river to a pre-chosen perch (which may be the rigging of a moored cruise boat), where the fish is gulped down whole and headfirst. As if its attention-drawing behavior were not enough, the kingfisher's call is distinctive, and surprising. From a bird with such a generously proportioned bill I was expecting a loud, deep call. The actual call is a rather high-pitched series of abrupt *keeks*.

The waters of the Red Sea get colder during winter yet never drop below the temperature critical for the survival of the coral, around 18°C. Even so, some divers claim that reef life slows down a little through the winter months. Not so for some of the big fish. December to February is the mating season for certain species of sharks, including the hammerhead. The name is purely descriptive, as the eyes are set on thick stalks that project either side of the head, giving it a hammer-shaped outline from above. The purpose of this is not altogether clear, but it may improve maneuverability or aid in sensory perception. Whatever, it looks weird. Several species of hammerhead occur in the Red Sea, but the most likely to be seen is the Scalloped Hammerhead. Despite a fearsome reputation, probably due in large part to their unusual appearance, this species is not known to attack humans. However, sharks during their mating season are reported to be more aggressive, so proper caution should be observed.

This time of year also marks the mating season of the Mantas. Most divers in the Red Sea are familiar with the Blue-spotted Ray. As in most rays, the body is extremely flattened. The tail has developed into a slender filament, in this species bearing a sting, and the ray moves by undulations of its body. The

Blue-spotted Ray is a bottom-dweller, olive-brown and covered in blue spots. It is about a meter across the 'wings.' The Manta is similar, but is black above and white below, up to seven meters across and 2,000 kilograms in weight. It is also called the Devil Ray because of the two fleshy protuberances or 'horns' on either side of its mouth. Despite its intimidating name, the Manta is a peaceable beast that feeds on plankton. The horns help channel the plankton into the fish's mouth, where the water is siphoned off, leaving a planktonic soup. In January and February the Mantas gather in small schools and head north, possibly to take advantage of the thick plankton blooms that appear at this time of year. Encountering a group of these awesome fish must be like coming across an unexpectedly waterbound squadron of light aircraft. The aircraft analogy is especially appropriate, for Mantas do occasionally take to the air. They leap out of the water and then crash back in. Many would like to think that this behavior, known as breaching, is sheer high spirits, but the more sober-minded scientists argue for some territorial explanation or behavior in competition for mates. The louder the bang the better the chance of success.

Passion is not confined solely to the water. While for us humans Egypt may be locked in winter's cool, though hardly icy, grip, come January the thoughts of the country's three foxes are turning to spring. The Fennec, already encountered hunting for small rodents and insects in spring, is the smallest. In winter the males belt out their mating howl, a mournful, quavering cry, deepening toward the end. Fennecs mate for life, probably an adaptation to life in inhospitable desert regions that can only support a very thin distribution. The female will give birth to up to five cubs sometime in April or May to take advantage of the plentiful spring food supplies.

Much more likely to be seen is the Red Fox. This is the same fox found throughout Europe and North America, though in Egypt it is generally rather gray, slender, and—frankly—scruffy. The white tip to the tail, or brush, however, is retained. In December and January, when the females come on heat, the Red Foxes along the desert edge can become very bold, and several

Red Fox. *Despite its name, there is little red about the race of Red Fox found in Egypt—it is more a dingy beige. However, it is one of the most easily observed mammals here.*

can be seen at a time. Invariably this group will be a single female accompanied by several males vying for her attention. I have found myself on horseback in the desert surrounded by up to five Red Foxes at close quarters. Unlike its European cousins, the Red Fox in Egypt does not seem to be an urban mammal, probably because of competition from feral dogs. However, they hold their own in the countryside, often living in dens in the desert and descending to the farmland at night to prey on rodents, small vertebrates, and anything else that might turn up. Interestingly, they seem to thrive in the same desert areas that support large packs of feral dogs. Perhaps in the country there is more to go round. During the rest of the year, Red Foxes may be seen but are less approachable. Watch out for one at the Sun Temple near Abu Sir with a bald tail! The third Egyptian fox is Rüppell's Sand Fox, a similar but smaller species, more restricted to the desert areas.

The thoughts of Egypt's three species of smaller wild cats have also turned toward spring. For all three—the Sand Cat, the Swamp Cat, and the Wild Cat—January courtship will be their one and only social event of the year. It will be a brief, yowling affair, after which the pair will part, the male to seek new partners, the female to prepare for a March/April motherhood. Of these three cats, by far the rarest is the diminutive Sand Cat. With flared paws, flattened head, and creamy coat, it is ideally adapted to its haunts in areas of remote sandy desert. The coat is relieved by bold markings on the forelegs especially. However, when the cat hunts these are concealed so as not to break the camouflage. An elusive species, its only confirmed sightings in the 1990s have

Sand Cat. *Rare and beautiful, the Sand Cat is one of Egypt's most elusive mammals. There have been only two confirmed records from Egypt in the 1990s.*

been of two individuals shot by hunters. One ignominiously and illegally ended up in the bazaar at Kerdasa, tastelessly stuffed and displayed amongst the more usual foxes and jackals.

More common, but really only marginally more likely to be seen, are the Swamp (or Jungle) Cat and the Wild Cat. The race of the latter found in Egypt, around the desert margins, is also known as the Kaffir Cat and is thought to be the ancestor of the domestic feline so familiar today. Although revered by the ancient Egyptians as the living image of the goddess Bastet, the animal is now thought to have first been domesticated in Mesopotamia some 8,000 years ago. The first representations of a domestic cat do not appear in Egypt until the New Kingdom, around 1450 BC. There are earlier pictures in tomb paintings and friezes of what are clearly cats, tabby-like cats at that. A beautiful example is the cat hunting waterfowl in the tomb of Nebamun at Thebes. Very successful it is too, with a white wagtail in its back paws, a shrike in one front paw and a duck of some kind in its other. The thing is that Wild Cats do look very like domestic tabbies, and there is no evidence to prove that this is a picture of a domestic cat. It is only in the New Kingdom that cats are portrayed in association with other tame and domestic animals, being fed, or tethered to chair legs. The same problem occurs today. The Kaffir Cat looks very like a rather pale, large tabby. Perhaps the best identification feature is the tail. It is shorter and thicker than that of most domestic cats and always ends in a black tip. Ultimately, though, if you get close enough to the cat to see its tail, it is highly unlikely to be a Kaffir Cat.

Bastet, the cat goddess, was worshiped at Bubastis, near the modern town of Zagazig. As a deity she rose to prominence during the Libyan dynasty some 2,900 years ago, when the Egyptian capital was moved to Bubastis for a spell. Bastet was the goddess of good fortune and joy, although she had originally been associated with the lion-headed goddess Sekhmet, who was responsible for disease and death, as well as protecting the pharaoh. There were great temples at Bubastis, but little remains today. It is a field of dry mud wasteland bordered by overcrowded blocks of flats and looking like a First World War

***Kaffir Cat**. This wild cat is the probable ancestor of the domestic cat, but mystery shrouds exactly when and where this domestication took place.*

battlefield put in the sun to dry. The trenches, though, are collapsed catacombs, for just as the ancient Egyptians mummified millions of ibises to venerate the God Thoth, they did the same with cats for Bastet. So many were found when the site was excavated in the last century that the archaeologists had no idea what to do with the things. Suggestions included using them as ship's ballast or grinding them down for fertilizer. The glut led to appalling wastage, and few of the mummies sent abroad survive today.

Not all the mummified cats were Kaffir Cats. Some were not cats at all but the bones of other animals, including humans, wrapped up and passed off as cats—possibly some priestly racket. But of those that were genuine cats, a few were not Kaffir Cats but Swamp Cats. Some have suggested that this is evidence that the Swamp Cat is mixed up somewhere in the genetic make-up of the domestic cat, but this is unlikely. The Swamp Cat is a distinct species, larger, longer-legged, and much shorter-tailed than the Kaffir Cat. Interbreeding has never been proven in the wild. Today, the Swamp Cat is still found in Egypt, though like its cousin, it is a reclusive animal. It is found in damper areas, including reedbeds and marshes.

Those familiar with the macho antler-clashing of frisky stags might be surprised to find that the courting of the smaller cats is also known as rutting. Like any specialist field, natural history has its own curious terminology, but I like to think that it is more picturesque than most, and nowhere more so than in the collective noun. Admittedly, at the level of sheep it is pretty dull. Everyone's heard of a flock. But how many Egyptologists realize that the six Meidum Geese, that most famous of Old Kingdom friezes now housed in the Egyptian Museum, represent a gaggle but that if they were to take flight they would become a skein? To see a single Crested Lark is hardly a memorable experience. They are the dull, streaky birds unconsciously familiar to out-of-town motorists as the brown birds that fly up from the road as the car is almost on top of them. Yet in a group these dowdy songsmiths become an exaltation. An exaltation of larks.

Other examples abound. Most days, for instance, I can look out of my apartment and see a murder. I have seen murders in Giza, Zamalek, Garden

City, Dokki, and most other areas of Cairo, yes, even Maadi. The mundanity of all of this only becomes apparent when one realizes that a 'murder' is a group of crows, in Cairo's case Hooded Crows. More pleasingly, a flock of Goldfinches is known as a charm. Now common in gardens throughout the Delta, this sparrow-sized bird with startling yellow wingbars is attracted by ripe seedheads, especially those of thistle species. The adult is otherwise a warm, pale brown, with a striking black and white head and scarlet face mark. The Goldfinch is something of a success story. Having become extinct as a breeding species in the Delta earlier this century, it has recolonized Cairo and its environs to the extent that it is now a frequent nester. In winter the resident birds are joined by 'charms' fleeing Europe's colder weather. Another, duller finch that may also be seen in winter flocks in the farmland is the Greenfinch. Slightly larger than the Goldfinch, it too has yellow wing patches but is otherwise dull green.

January 1992 was marked by a series of anniversaries. The New Year saw the 412th anniversary of the last report of a Hippopotamus in Cairo. It was 101 years since the last reliable sighting of a Sacred Ibis in the country and 80 years since the death of Egypt's last Wild Boar. The latter passed away rather ignominiously in Cairo Zoo. The Hippo and the Sacred Ibis were probably both victims of habitat change, but the Wild Boar's demise was more likely due to overhunting. Unfortunately, things do not change. January 1992 also marked the arrival of a group of Saudi hunters given permits to hunt in the desert south of Mersa Matruh. The potential carnage was enormous. This was not a couple of duck hunters off on a weekend jaunt but, according to the Egyptian Gazette, a convoy of 23 coaches on a three-week hunting trip. The press announcement stressed that they were only permitted to hunt on the condition that they use trained falcons rather than rifles. The traditional quarry for Gulf falconers is the Houbara Bustard, and ominously the very region the hunters were heading for was the last Egyptian stronghold of the Houbara. It is a handsome, stout gamebird, the size of a large, heavily-built chicken, with a striking black and white neck pattern. Having hunted the bird to virtual extinction in their own area, the Gulf hunters must now travel to annihilate populations in neighboring regions. I am not sure of the collective noun for the Houbara but whatever it is, it must rapidly be coming redundant.

The stretch of desert in the northwesternmost corner of Egypt is also home to a couple of other birds found nowhere else in Egypt. The distributions of the Barbary Partridge and Dupont's Lark both stretch across North Africa and just into Egypt across the Libyan border. The former is a typical partridge, chunky and short-winged, with a gray head, chestnut breast band spotted white and black, and maroon and cream striped flanks. The latter is not unlike a Hoopoe Lark in shape but has a shorter, less markedly curved bill and a more uniform rufous-brown plumage. The status of both birds is at present uncertain because access to this part of Egypt, along the Libyan border, is difficult.

The Western Desert south of the Mediterranean used to support large herds of gazelle, but alas no more. As little as twenty years ago, gazelle were seen regularly as close to Alexandria as al-'Alamein. Now there are very few reports. Hunting and habitat disturbance have ensured their demise. There are three species of gazelle in Egypt. The Arabian Gazelle is possibly still found in northeastern Sinai. The Slender-horned or White Gazelle is extremely rare but used to be found in the Western Desert in the Qattara depression and south of the Fayoum. Also known as Loder's Gazelle or the Rhim, it probably has more names than individuals left here. The only species of which significant numbers still remain is the Dorcas Gazelle. It, too, is disappearing but can still be found in the remoter wadis of the Western and Eastern deserts.

The Dorcas Gazelle is a graceful animal some 60 centimeters high at the shoulder and weighing around 15 to 20 kilos. It is largely pale brown, lighter beneath, with a dark band along the flanks that fades somewhat in the summer. The males have lyre-shaped horns. A whitish stripe runs from each horn base to the muzzle. In Egypt, Dorcas Gazelles mate in late summer and give birth some six months later in late winter. The good news as far as the gazelle is concerned is that courtesy of humans its main natural predators, the cheetah and the leopard, have been hunted to probable extinction in Egypt. The bad news is that the Dorcas Gazelle is in danger of going the same way.

Now extremely wary, it has retreated to the remoter areas of the desert, where it is still persecuted. Stuffed gazelle are still sold regularly at the taxidermy shops in Kerdasa. For those with a sense of the especially tasteless, it is possible to buy a stuffed mother being suckled by a stuffed calf. One of the shops boasts a picture of a jeep draped with the corpses of a dozen or so gazelle, the result of one hunting trip.

Evidence of gazelles is found far more often than the animals themselves. Their tracks can be identified in the desert when they are nowhere to be seen. Other information is anecdotal. I once caught a service taxi from Quseir on the Red Sea coast across the Eastern Desert to Qena, north of Luxor. The scenery was wild, craggy, mountainous, and austere, the road winding its way through narrow ravines, their bases sprinkled with desert scrub, evidence of subsurface water. In the midst of this wilderness was a tea shop, a small stone-and-cement hostelry that seemed barely distinguishable from the rock that towered over it. On the wall of the tea shop was the most beautiful stylized painting of a gazelle in mid leap. Over a wickedly strong, wickedly sweet brew, I asked the proprietor whether he ever saw gazelle. He smiled at my naiveté. He used to, but not now the road was built. The gazelles had gone to wadis higher and more remote. He had not seen them for a long, long time.

Winter trips down to the Fayoum always seem to produce surprises. It is, of course, always pleasant to renew old acquaintances, the flamingos especially, but there always seems to be something new down there. Two especially welcome guests last winter were the Spoonbills and the Avocets. The Spoonbill is one of those rare birds for whom the name is extremely apt. These large, all-white, egret-like birds have a bill shaped just like a spoon. They use this curious appendage to sift through the mud in the lake shallows for small fish and crustaceans. Although Spoonbills look like egrets they are in fact much more closely related to the ibises, despite the fact that ibises have long, slender downcurved bills. In flight, Spoonbills can be told from egrets even when the 'spoon' is not clear. All herons and egrets fly with the neck tucked back against the shoulders. Spoonbills fly with the neck outstretched.

To my mind the Avocet is the most elegant of birds. Strikingly marked in black and white, it is a typical wader with long, gray-blue legs. However, as with the Spoonbill, its most distinctive feature is its bill. The Avocet's bill is slender, black, and sharply upturned. It wades through the water sweeping its recurved bill from side to side through the mud, picking up worms and other invertebrates. It can often be seen in company with another elegant wader, the Black-tailed Godwit. The godwit has a very long, dead-straight bill. While the Avocet advances through the water sweeping to and fro, the godwit strides forward prodding and stabbing for prey more deeply hidden. Thus the two birds can cover the same patch of mud without competing—and what a striking duo they make.

Avocets, godwits and Spoonbills are all very distinctive and easy to both spot and identify. However, Lake Qarun and many wetland areas of the Delta support other waders that are memorable not only for their numbers but for their apparent sameness. Redshanks, Greenshanks, Common, Green, Wood, and Marsh sandpipers, Dunlins, Little Stints, Sandplovers, Little Ringed and Ringed plovers, and Ruff can all be found but can be frustratingly difficult to identify. In winter many of them loose their distinctive summer plumages, worn on their northern breeding grounds, and don duller, more uniform winter garb. Binoculars and a good guidebook are necessary to differentiate many of them, and even then it can be difficult. However, as winter ends and the birds prepare to fly north they start to assume their more colorful breeding plumage.

One winter I had been repeatedly frustrated by a small wader I kept seeing on the shingle beaches of the southern shore of Qarun. It was a small, fairly short-legged bird, gray-brown above, white below, with a faint breastband. The rather stout bill, black and slightly swollen at the tip, identified it as a plover. Further observation narrowed it down to a sandplover, but Lesser or Greater? I wanted it to be Lesser because that is much rarer in Egypt, but I found it impossible to be certain. My field guide was not much help, saying merely that the two were 'distinguishable in winter only by size.' The Greater

***Spoonbills** and **Avocets** (foreground). Both these species feed in the muddy shallows of wetland areas, here in Lake Qarun. Their strikingly different bills reflect their different methods of feeding.*

is only three centimeters larger than the Lesser. When you have nothing to compare with that will stay still, and the thing is a hundred meters away, three centimeters is not very much. However, after repeated observation in the same area I came reluctantly to the conclusion that my plover was the more common Greater. But I could not be certain. Then I saw the bird in late winter donning its breeding dress. It assumed a black 'bandit' mask, a contrasting white forehead, and a narrow chestnut breast band. The latter confirmed everything. The Lesser Sandplover has a rufous-brown breast, almost too broad to be called a band. My disappointment with it being the more common bird was tempered by the satisfaction of having identified it correctly through the winter.

Other birds, too, undergo changes as they prepare to head north. Two species of grebe winter in Egypt (a third, the Little Grebe, is resident). The Great Crested and Black-necked grebes are both common on the larger lakes. Both are dull gray above and white below, but the first is much larger than the second. They swim lower in the water than ducks and have sharp, pointed bills. For most of the winter the Great Crested Grebe bears nothing that really deserves the name crest, and the Black-necked Grebe sports a neck that is not black, rather pale gray. Dowdy and unassuming, they often go unnoticed. However, prior to migration they undergo something of a transformation. It becomes apparent how they both earned their names. The Great-crested sports a deep brown, double-horned crest and dramatic orange cheek plumes that it will raise in an impressive ruff on its northern courting grounds. For its part, the Black-necked gains a black neck and bright orange eartufts. These colorful summer plumages are only seen briefly in Egypt, for once assumed the birds migrate.

Bats apart, Weasels are the most likely of Egypt's mammals to be seen by the casual observer. However, the diminutive yet rapacious little predator seen scampering beneath the bellies of parked cars is not the only weasel in Egypt. There are two others. One, the Zorilla, is confined to the southeasternmost corner of the country. The second is the Striped Weasel, to be found along the western margins of the Delta across toWadi Natrun and beyond. Larger than our familiar urban weasel, it is also more strikingly colored and more heavily furred. The underside is uniform dull black. The upperparts are pale cream striped lengthwise with four to five dark stripes. The shaggy tail is either predominantly black or white, and there is a distinctive pale ring around the face just behind the eye. The Striped Weasel is rarely seen, as it is nocturnal. Being largely black and white, comparison with the skunks of America is tempting. For once, appearance does not deceive. Skunks and Striped Weasels share the same highly offensive defense mechanism. The striking patterning serves as a warning to potential predators that the Striped Weasel should not be messed with. On being threatened, the weasel has the good grace to warn the assailant that something rather unpleasant is imminent. It turns round, raises its tail and presents its posterior to the threat. If the threat persists then the enraged mustelid lets fly the ample contents of its stink glands in a nauseating, foul-smelling spray. In February and March the female Striped Weasels will be giving birth to litters of two to three kittens in rocky dens. Having smelled a Striped Weasel for myself, I consider rocky dens an excellent place for them.

What with mating sharks, maternal weasels, and myriads of birds slipping into their courtship plumage, changes are in the air. The second half of February sees the first birds passing through on their way north to Europe, the forefront of the great migration. As the air warms the insects increase, the bats come out, and the Kestrels outside my building start screaming at each other once more. Faunal shufflings are once again afoot. Spring is on its way.

One of the aims of this book has been to share with the reader the variety of wildlife that Egypt has to offer. While it is hoped that interest in Egypt's wildlife may be sparked off or rekindled, it has been beyond the scope of this book to provide a comprehensive guide to Egyptian wildlife. However, humans are by nature curious beings. For those who wish to go further and identify the creatures they may see during their time in Egypt, it is hoped that the following guide to the field guides will be of help. Unfortunately, only the birds and the marine life of the Red Sea are covered in any detail by the currently available literature. For the other groups of animals, coverage is much scantier.

The Birds

There is now a plethora of books that cover the birds of Egypt. The problem is local availability. For most visitors the *Common Birds of Egypt* by Bertel Bruun and Sherif Baha el Din (The American University in Cairo Press, Revised Edition 1990) will prove sufficient. It covers those species most likely to be seen, is slim enough to slip into a travel bag almost unnoticed, and has the distinct advantage, like all the books published by the AUC Press, of being widely available in Egypt. It also contains a complete checklist to the birds of Egypt. For the more serious birdwatcher by far the best guide is *The Birds of Britain and Europe with North Africa and the Middle East* by H. Heinzel, R. Fitter, and J. Parslow (Collins, fourth edition reprinted and revised 1984). The plates are sometimes misleading, but it is the only guide that covers all the birds of the country in a single volume. The alternative is *Birds of the Middle*

East and North Africa by P.A.D. Hollom et al. (Collins, 1988). This book covers all the birds found in Egypt but only illustrates those species or subspecies not covered in its companion volume *A Field Guide to the Birds of Britain and Europe* by R. Peterson, G. Mountfort, and P.A.D. Hollom (Collins, fourth edition 1983). Thus the birdwatcher, already weighed down with notebook and binoculars, has to cart round two books. Copies of all these titles are available only spasmodically in Egypt.

Not available here, and only for the ultra-enthusiast with a lot of cash to spare, is *Birds of Egypt*, edited by S. Goodman and P.L. Meininger (Oxford University Press, 1988), the current definitive text on Egypt's birds.

A historical perspective of Egypt's birdlife can be obtained from *The Birds of Ancient Egypt* by P. Houlihan with the collaboration of S. Goodman (The American University in Cairo Press, 1988). Every species of bird appearing in pharaonic art or hieroglyphs is covered, and is described and illustrated by photographs or line drawings of the original representations. Perhaps most interesting are those birds represented but no longer found in modern Egypt.

The Mammals

There is no widely available book specifically on the mammals of Egypt. The definitive work is *The Contemporary Land Mammals of Egypt (including Sinai)* by D. Osborn and I. Helmy (Field Museum of Natural History, 1980), but this is unavailable in Egypt and nigh on impossible to find elsewhere. The larger mammals of Egypt, including all the carnivores, are covered by *A Field Guide to the Mammals of Africa including Madagascar* by T. Haltenorth and H. Diller (Collins, 1980). Most species are illustrated in color and the text is concise and informative, covering habits and distribution as well as appearance. The smaller mammals are harder to find information on. *The Mammals of Britain and Europe* by G. Corbet and D. Ovenden (Collins) covers those European species whose range extends into Egypt, but I have never seen it in an Egyptian bookstore. Of the above, only the last covers sea mammals (dolphins, porpoises, and whales).

Reptiles and Amphibians

These creatures suffer more on the book front than even the mammals. I still refer to *Zoology of Egypt: Volume 1* by J. Anderson (Bernard Quaritch, 1898) for identification, but copies are few and far between, to put it mildly. H. Marx compiled a *Checklist of the Reptiles and Amphibians of Egypt* (Special Publication United States Naval Medical Research Unit Number 3) in 1968, but it too is dated and offers little help in identification. The beautifully illustrated *Field Guide to the Reptiles and Amphibians of Britain and Europe* by E.N. Arnold, J.A. Burton, and D.W. Ovenden (Collins, 1978) covers a number of species found in Egypt and should help the user to at least narrow identification down (for example, is this snake a sand boa or a viper). It covers all the sea turtles recorded from Egyptian waters. As with all the Collins guides, availability in Egypt is erratic.

Fish and Marine Life

There are now several guides available to the fish of the Red Sea, none of which is definitive and all of which have their pros and cons. My personal favorite is *Red Sea Fishes* by Helmut Debelius (Verlag Stephanie Naglschmid, 1987). It has clear photographs of a wide selection of Red Sea reef fish, many taken by the author, in their natural habitat. The drawback is the binding. Heavy use and it falls apart. *The Diver's Guide to Red Sea Reef Fishes* by J.E. Randall (Immel Publishing, 1982) also illustrates a wide range of reef fishes, but the fish are photographed dead and out of water. Many look strikingly different when seen underwater. The guide has the advantage of being printed on waterproof paper. The *Red Sea Fish Guide* by R. Deuvletian (1987) is a cheaper volume with rather inferior photos, but it is ideal for the mildly curious snorkeler. For those who want to know more about the coral reef ecosystem as well as the fish that inhabit it then the lavishly illustrated *The Red Sea in Egypt, Volume 1: The Fishes*, written, illustrated, and published by F. Atiya (1991) is well worth the extra expense of a hardback book.

As far as the invertebrate inhabitants of the reef go, the best coverage is that given in *Red Sea Invertebrates* by P. Vine (Immel Publishing, 1986).

All the above are readily available in Cairo from most hotel bookstores and, at a price, in Sharm al-Shaykh and Hurghada. The same is not true of books relating to Egypt's other coast. For the Mediterranean, the excellent *Hamlyn Guide to the Mediterranean Sea* is unfortunately out of print. The best alternative is the *Guide to Inshore Marine Life* by D. Erwin and B. Picton (Immel Publishing), but I have not seen it on sale in Egypt.

The fish of the Nile are covered, perhaps surprisingly, by *Fish and Fishing in Ancient Egypt* by D. Brewer and R. Friedman (The American University in Cairo Press, 1990). A companion volume to *The Birds of Ancient Egypt*, it uses much the same format, giving information about each species of fish and illustrating each entry with photographs of their representations in pharaonic art.

Other Animals

Books on the insects, spiders, scorpions, and other invertebrates found in Egypt are few and far between. *A Field Guide to the Insects of Britain and North West Europe* by M. Chinnery (Collins, 1976) is far more useful than its name suggests. It should enable the user to at least narrow down the family to which the insect in want of identification belongs. It includes plates of a number of Mediterranean species that occur in Egypt, especially the hawk moths. Dragonflies are almost covered by *A Field Guide to the Dragonflies of Britain, Europe and North Africa* by J. d'Aguilar, J.-L. Dommanget, and R. Prechac (Collins, 1986). I say 'almost' because although Egypt is in North Africa, and the title implies that North Africa is covered, this is actually only North Africa from Libya west. However, most Egyptian species are described and illustrated and the book is available in Cairo—I bought my copy downtown. Happily, butterflies are fully and comprehensively covered. *The Butterflies of Egypt* by T. B. Larsen (Apollo Books/The American University in Cairo Press, 1990) describes and illustrates every species of butterfly found in Egypt.

Plants

The fauna of any country cannot be divorced from its flora. Fortunately there are two books that describe most of the plants likely to be encountered in Egypt, though strictly desert species are not covered. Both are readily available in Cairo. *The Weed Flora of Egypt* by L. Boulos and M. N. el-Hadidi (The American University in Cairo Press, 1984) covers over 160 of the flowering plants and ferns found in the agricultural areas and in the towns and villages. Each plant is described and illustrated in black and white. Trees are dealt with by *The Street Trees of Egypt*, by M. N. el-Hadidi and L. Boulos (The American University in Cairo Press, Revised Edition 1988). It follows a similar format, describing and illustrating some 50 tree species.

INDEX

Illustrations are indicated by page numbers in bold type.